Specialists in mathematics education

Mathematics 1

PYP 1 Write-on Workbook

for use with

IB Primary Years Programme

Michael Haese Mark Humphries Philippa Sawyers

MATHEMATICS PYP 1

Michael Haese B.Sc.(Hons.), Ph.D.
Mark Humphries B.Sc.(Hons.)
Philippa Sawyers B.Ed.(Hons.)

Published by Haese Mathematics
152 Richmond Road, Marleston, SA 5033, AUSTRALIA
Telephone: +61 8 8210 4666
Email: info@haesemathematics.com
Web: www.haesemathematics.com

National Library of Australia Card Number & ISBN 978-1-922416-73-5

First Edition 2024

Cartoon artwork by James Hobbs, Yi-Tung Huang, Le Quynh Chi Ta, and Ngoc Vo.

Artwork by Hannah Coleman and Brian Houston.

Cover art by Le Quynh Chi Ta.

Computer software by Patrick French, Ben Hensley, Huda Kharrufa, Rachel Lee, and Han Zong Ng.

Audio recorded by Eloise Quinn-Valentine.

Production work by Hannah Coleman, Michael Mampusti, and Ngoc Vo.

Typeset in Australia by Charlotte Frost and Deanne Gallasch. Typeset in Times Roman 13.

Printed in China by Prolong Press Limited.

Acknowledgements: While every attempt has been made to trace and acknowledge copyright, the authors and publishers apologise for any accidental infringement where copyright has proved untraceable. They would be pleased to come to a suitable agreement with the rightful owner.

FOREWORD

Mathematics PYP 1 has been designed and written for the International Baccalaureate Primary Years Programme (IB PYP). The workbook and online interactive material provide an engaging and structured package, allowing students to explore and develop their confidence in Mathematics.

Our aim with this spiral bound write-on workbook is to provide the students and teachers with an alternative to photocopied worksheets, which in turn will give the students an organised record of their work throughout the year.

The book contains a variety of exercises ranging from basic to advanced, to cater for a range of student abilities and interests. The material is presented in a clear, easy-to-follow style to aid comprehension and retention, especially for English Language Learners. A Revision set for each chapter is provided at the end of the book.

Important information and key notes are highlighted, and worked examples provide clear instruction and relevant explanations. Discussions, Activities, and Puzzles are used throughout the chapters to develop understanding. All of these features can be viewed through Snowflake, our online digital textbook platform.

We have endeavoured to provide a stimulating workbook and accompanying interactive material. Our aim is to develop and encourage student understanding and to grow an appreciation and love for Mathematics.

We welcome your feedback:

Email: info@haesemathematics.com
Web: www.haesemathematics.com

ONLINE FEATURES

Each workbook comes with a 12 month subscription to the online edition and its range of interactive features. This can be accessed through the **SNOWFLAKE** online learning platform via a web browser or our offline viewer.

To activate your electronic workbook, please contact Haese Mathematics by emailling info@haesemathematics.com with your proof of purchase such as a copy of your receipt.

For general queries regarding **SNOWFLAKE** and online subscriptions:

- Visit our help page: https://snowflake.haesemathematics.com.au/help
- Contact Haese Mathematics: info@haesemathematics.com

WATCH LISTEN LEARN

Watch Listen Learn is an exciting feature of this book.

This icon in a worked example or exercise denotes an active online link.

Simply click the icon (or anywhere in the example or exercise box) to access the Watch Listen Learn with a teacher's voice explaining each concept.

Play any line as often as you like. See how the basic processes come alive using movement and colour on the screen.

CLICKABLE ICONS

You will see these icons throughout the workbook. Click the icon to access online content.

Printable Worksheets containing extra questions or support material.

Listening Activities.

Video Demonstrations providing instruction.

Activities and Games.

TABLE OF CONTENTS

CHAPTER 1: NUMBER

We use **numbers** to **count**.

Exercise 1

Practise your writing.

1 one

1 one

2 two

2 two

3 three

3 three

4 four

4 four

5 five

5 five

Exercise 2

Practise your writing.

6 six

6 six

7 seven

7 seven

8 eight

8 eight

9 nine

9 nine

10 ten

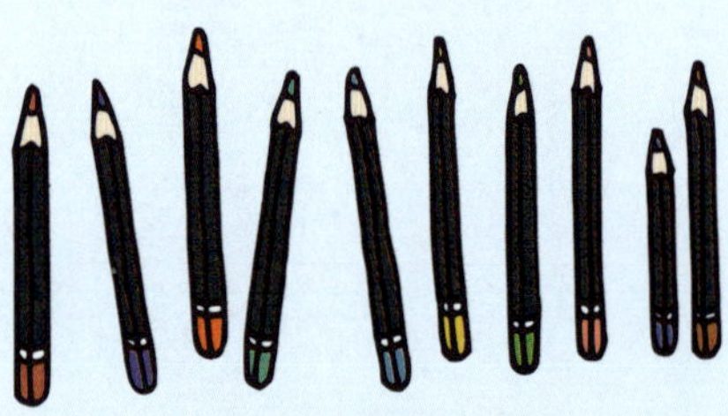

10 ten

Listening Activity

Click to hear how we *say* the numbers.

Activity

Count:

a trees ______ **b** lamps ______ **c** seats ______

d people ______ **e** birds ______ **f** balloons ______

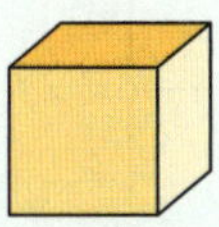

This block is a **unit**.

Small numbers can be drawn using unit blocks.

Exercise 3

Practise your writing. Copy each number.

1 one ______

2 two ______

3 three ______

4 four ______

5 five ______

6 six ______

7 seven ______

8 eight ______

9 nine ______

10 ten ______

Exercise 4

Complete:

a 1, 2, ______, 4, ______, ______, 7, ______, ______, 10

b one, ______, three, ______, five, ______, ______, ______, nine, ______

Exercise 5

Complete:

Diagram	*Numeral*	*Word*
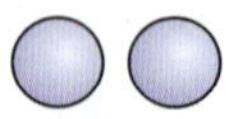	2	two
	5	
		seven
	3	
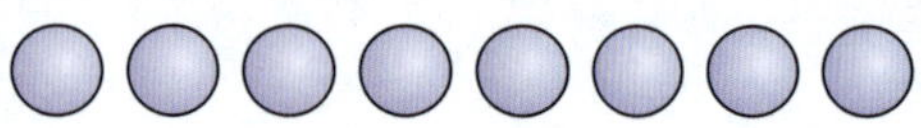		
		six
	4	

Activity

Count:

a

______ rabbits

b

______ butterflies

c

______ bees

d

______ kittens

Activity

Draw:

a 7 squares 

b 5 stars 

This block is a **ten**. 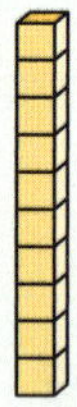 It represents 10 units. 

Exercise 6

Practise your writing.

20 twenty  20 twenty

30 thirty 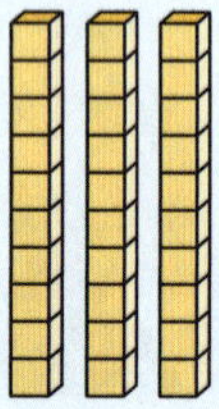 30 thirty

40 forty 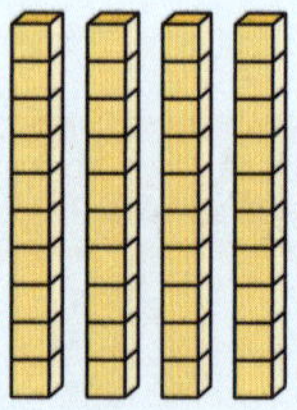 40 forty

50 fifty 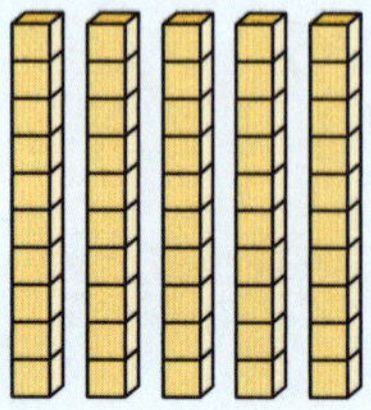 50 fifty

60 sixty 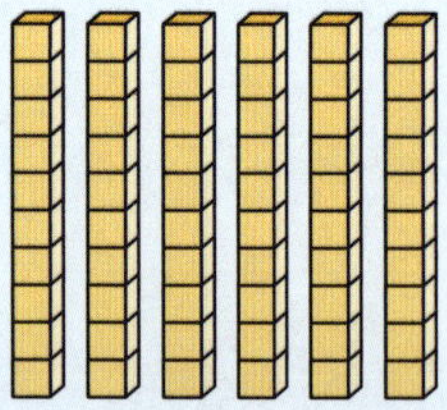 60 sixty

Exercise 7

Practise your writing.

70 seventy

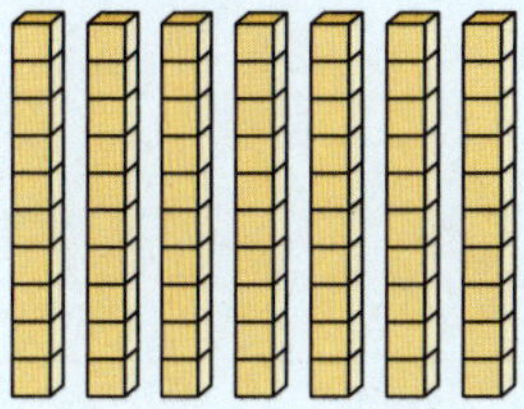

70 seventy

80 eighty

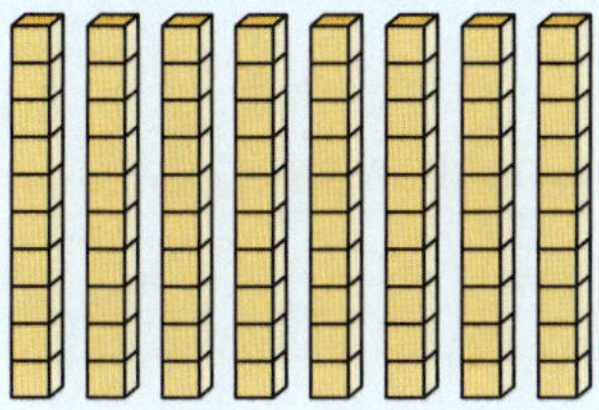

80 eighty

90 ninety

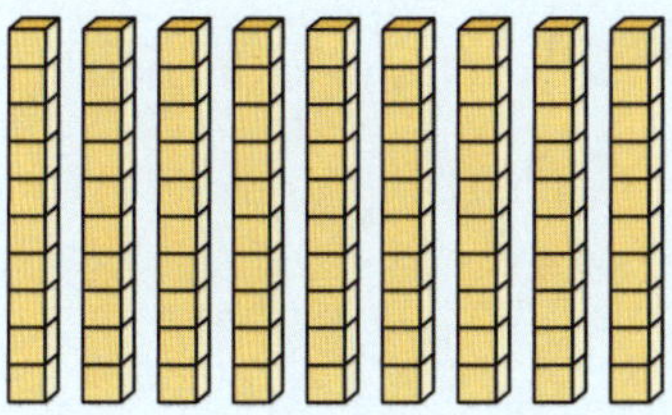

90 ninety

100 one hundred

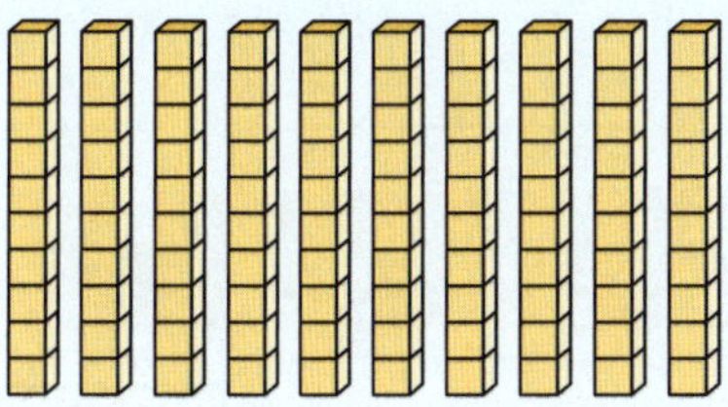

100 one hundred

Listening Activity

Click to hear how we *say* the numbers.

Exercise 8

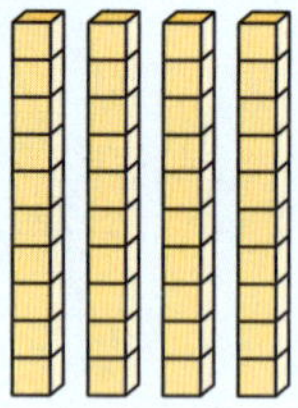

There are 4 ten blocks in 40.

a

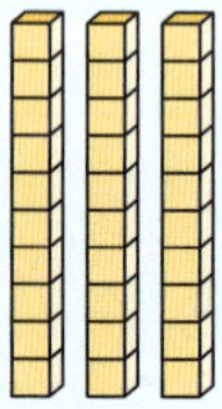

There are ______ ten blocks in 30.

b

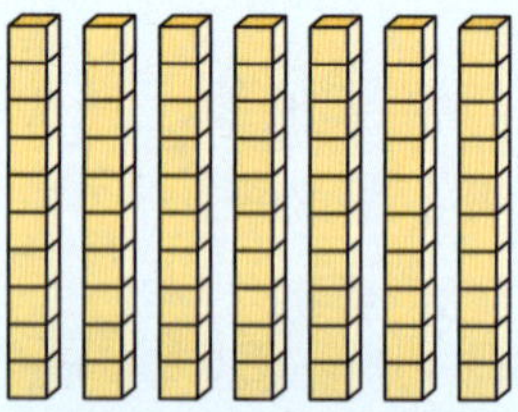

There are 7 ten blocks in ______ .

c

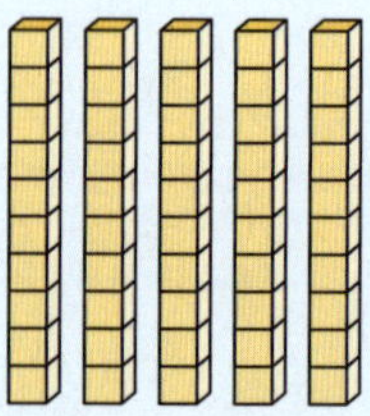

There are ______ ten blocks in ______ .

d

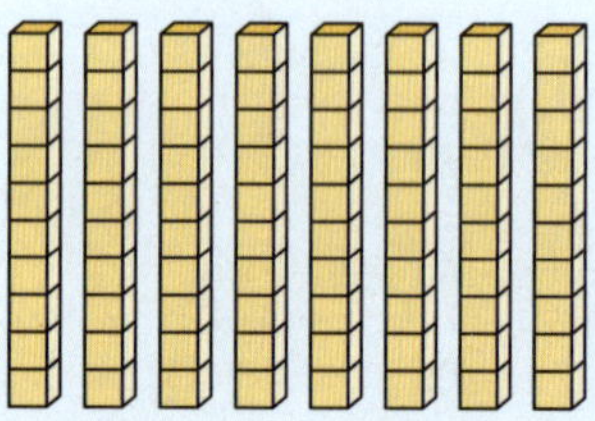

There are ______ ten blocks in ______ .

Listening Activity

Click and follow the instructions.

1 ____________________ **2** ____________________

3 ____________________ **4** ____________________

5 ____________________ **6** ____________________

7 ____________________ **8** ____________________

9 ____________________ **10** ____________________

11 ____________________ **12** ____________________

"equal to"

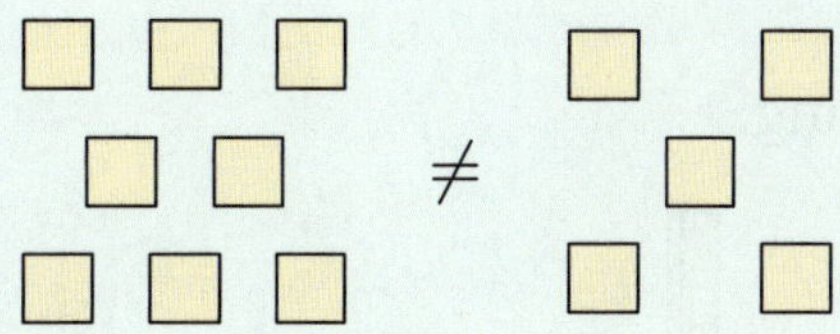

"not equal to"

Exercise 9

Practise your writing.

= equal to ≠ not equal to

___ ______________ ___ ______________

___ ______________ ___ ______________

Exercise 10

Use $=$ or $\neq$.

a

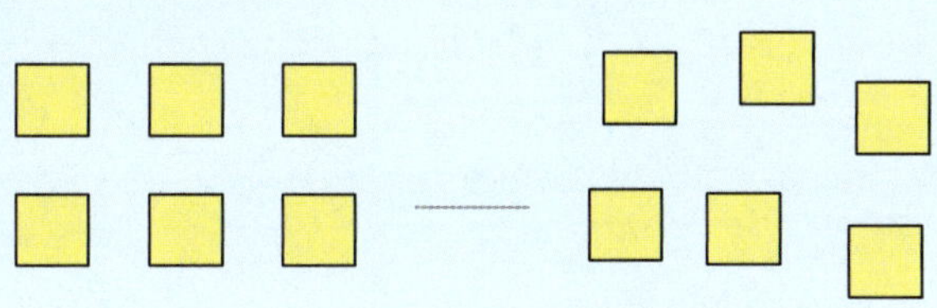

b

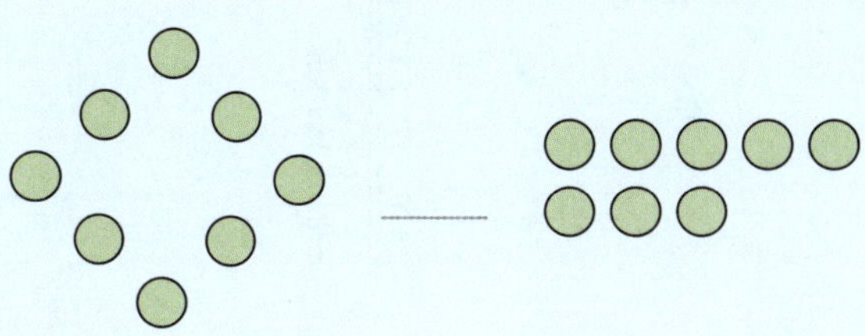

c

d

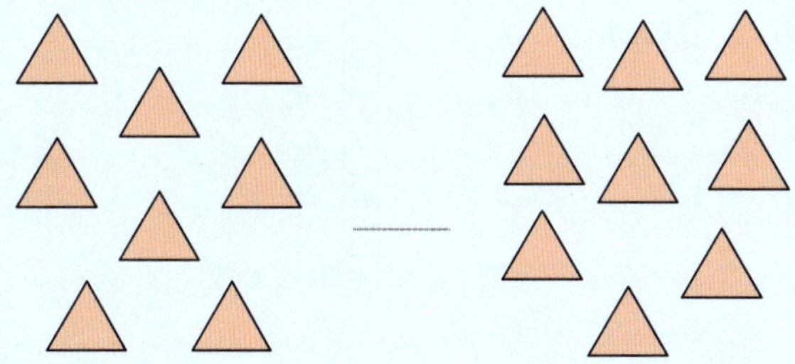

e

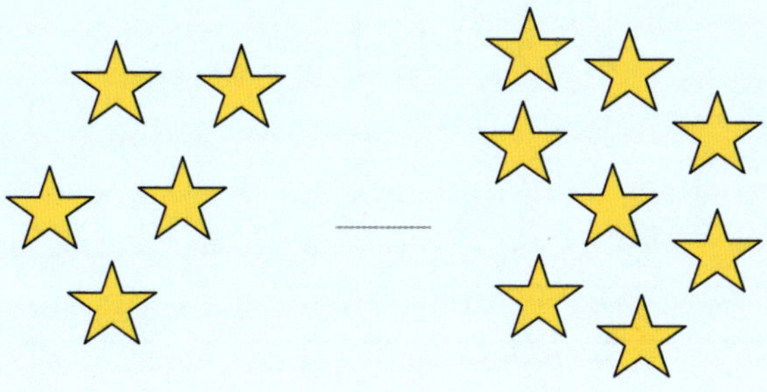

f

Exercise 11

Practise your writing.

11 eleven

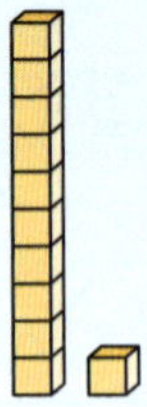

11 eleven

12 twelve

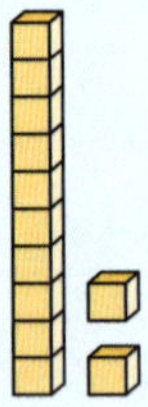

12 twelve

13 thirteen

13 thirteen

14 fourteen

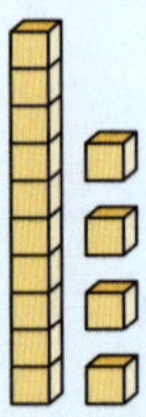

14 fourteen

15 fifteen

15 fifteen

16 sixteen

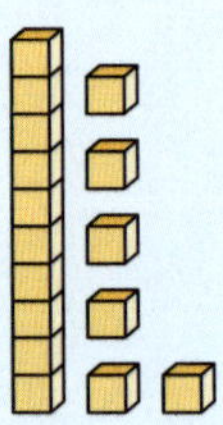

16 sixteen

Exercise 12

Practise your writing.

17 seventeen

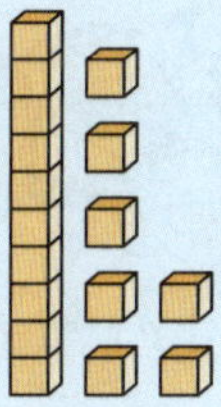

17 seventeen

18 eighteen

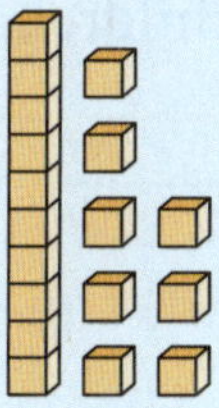

18 eighteen

19 nineteen

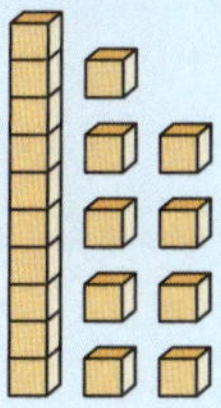

19 nineteen

20 twenty

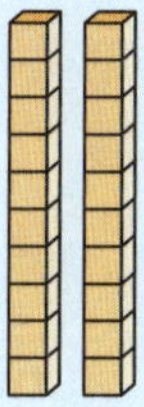

20 twenty

Listening Activity

Click to hear how we *say* the numbers.

Game

Click to practise your counting.

Activity

Count:

a 

______ eggs

b 

______ children

Exercise 13

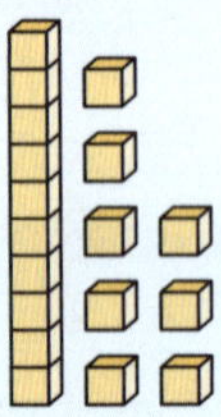

18 has 1 ten block and 8 unit blocks.

a 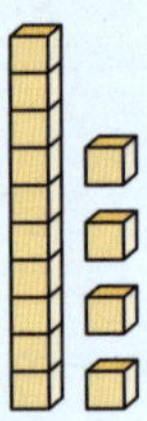

14 has 1 ten block and ______ unit blocks.

b 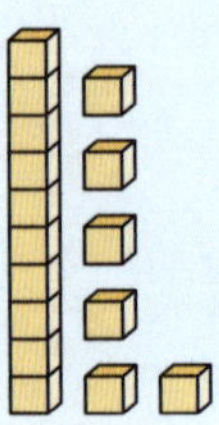

16 has ______ ten block and ______ unit blocks.

c 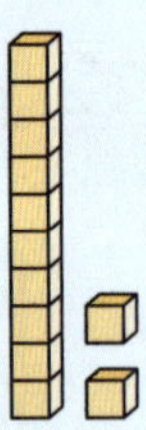

______ has ______ ten block and ______ unit blocks.

d 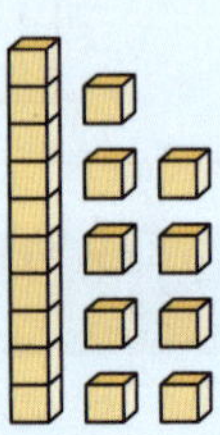

______ has ______ ten block and ______ unit blocks.

Activity

Colour the number shown:

4 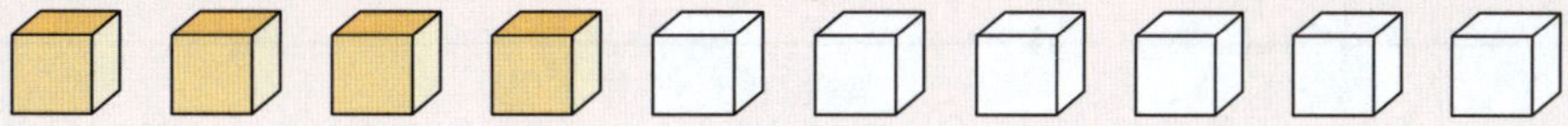

a 7 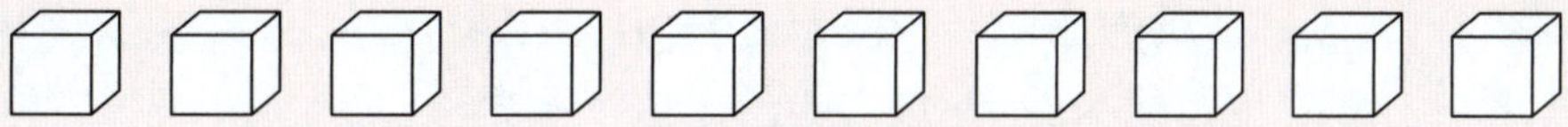

b 30 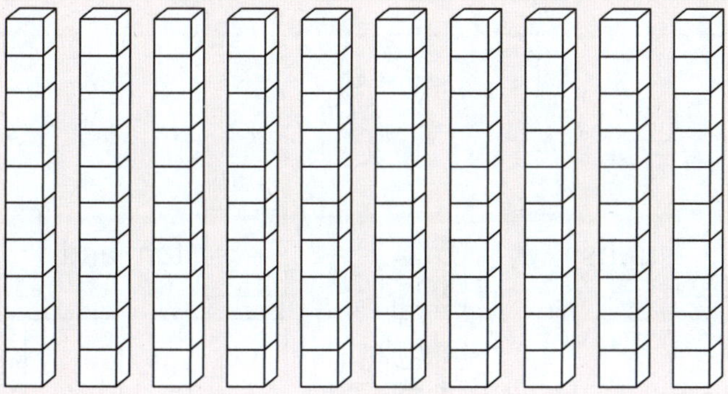

Exercise 14

Use $=$ or $\neq$.

a 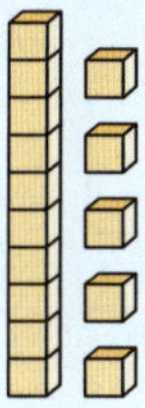 .... 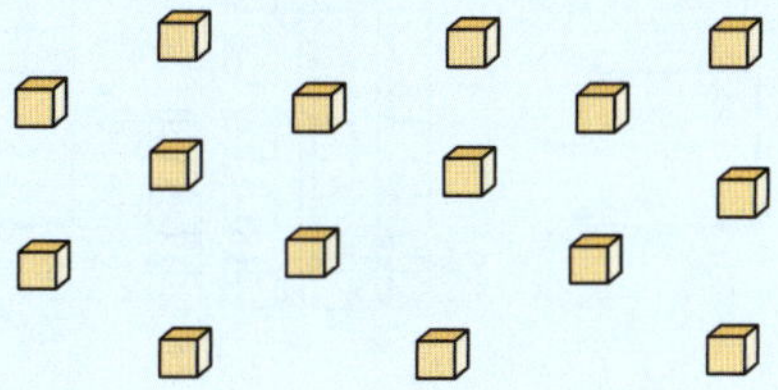

b 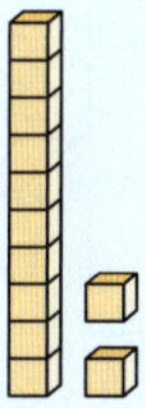 .... 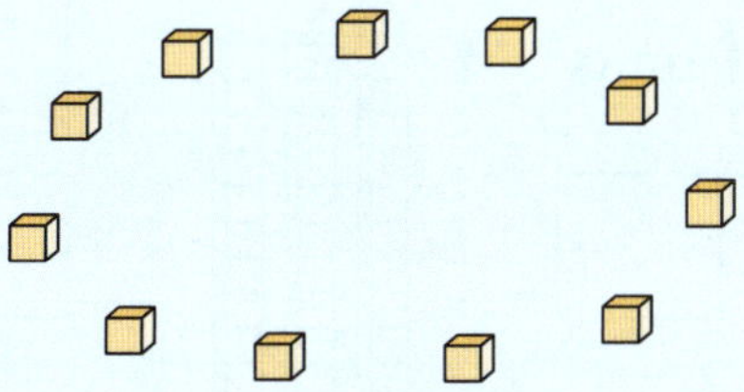

c 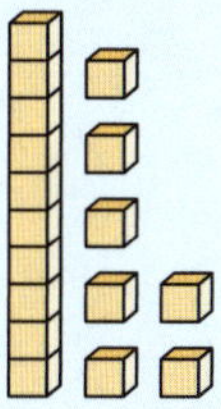 .... 

We write our numbers using **place values**.

Tens and **units** are place values.

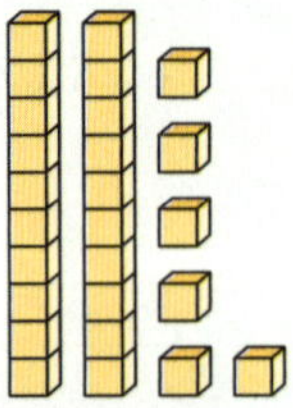

2 tens and 6 units

$$20 + 6 = 26$$

"twenty six"

Exercise 15

What numbers are shown?

a 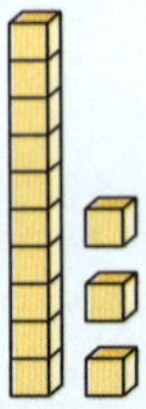

☐ ten and ☐ units

☐ + ☐ = ☐

b 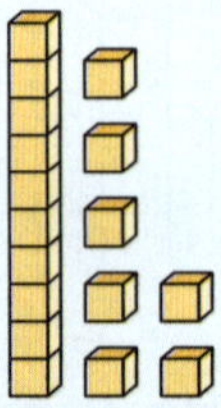

☐ ten and ☐ units

☐ + ☐ = ☐

c 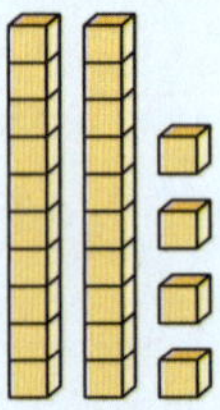

☐ tens and ☐ units

☐ + ☐ = ☐

d 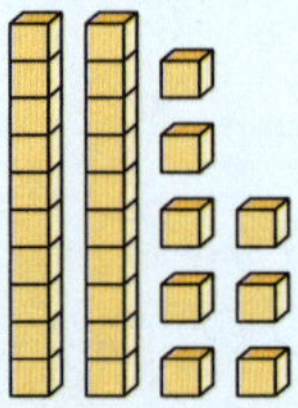

☐ tens and ☐ units

☐ + ☐ = ☐

e 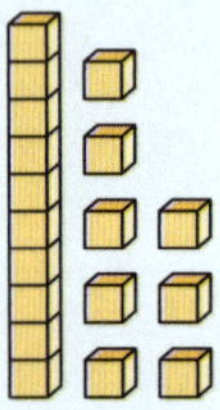

☐ ten and ☐ units

☐ + ☐ = ☐

f 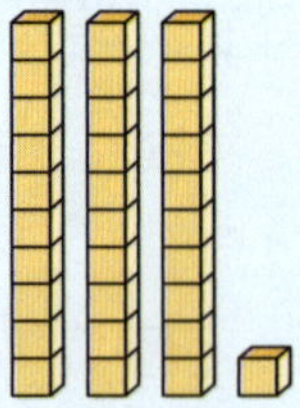

☐ tens and ☐ unit

☐ + ☐ = ☐

g 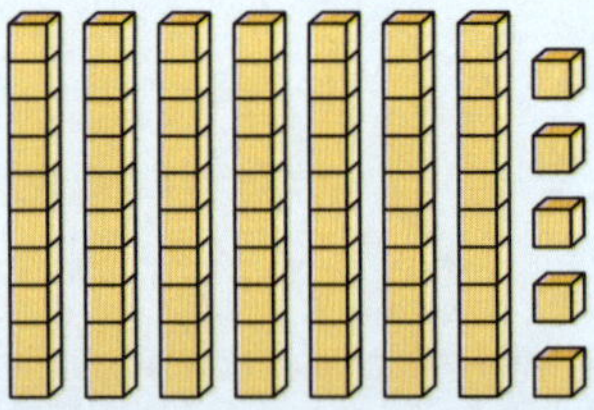

☐ tens and ☐ units

☐ + ☐ = ☐

0 is **zero**.

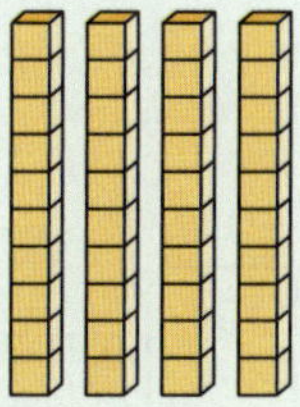

4 tens + 0 units = 40

"forty"

There are zero units.

Exercise 16

a

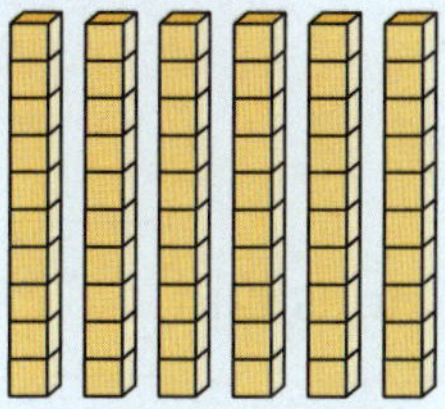

☐ tens + ☐ units = ☐

____

There are zero units.

b

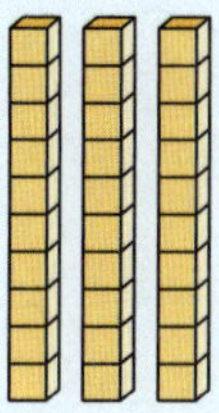

☐ tens + ☐ units = ☐

____

There are ______________ units.

c

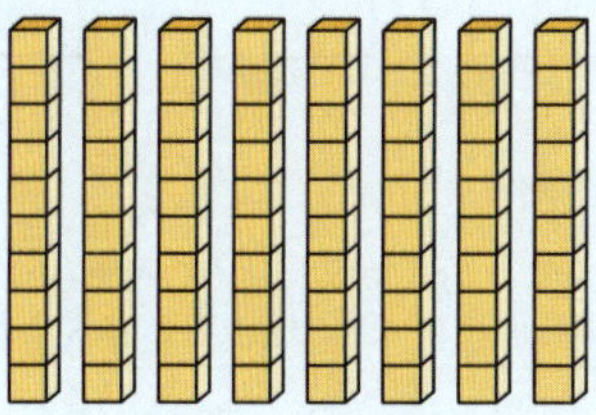

☐ tens + ☐ units = ☐

____

There are ______________ units.

Activity

Draw 16 marbles.

Colour 9 of the marbles red.

Exercise 17

What numbers are shown?

a 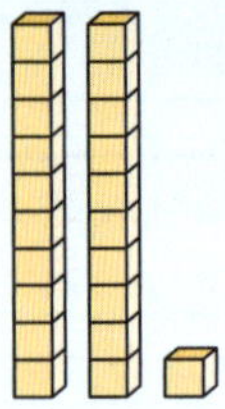

______ tens and ______ unit

______ + ______ = ______

______

b 

______ ten and ______ units

______ + ______ = ______

______

c 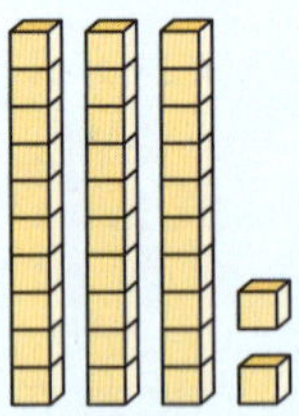

______ tens and ______ units

______ + ______ = ______

______

d 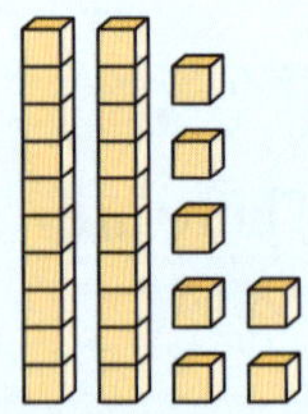

______ tens and ______ units

______ + ______ = ______

______

e 

______ ten and ______ units

______ + ______ = ______

______

f 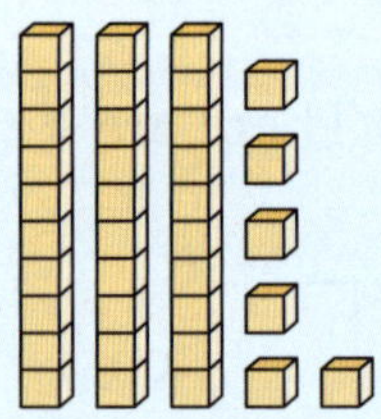

______ tens and ______ units

______ + ______ = ______

______

g 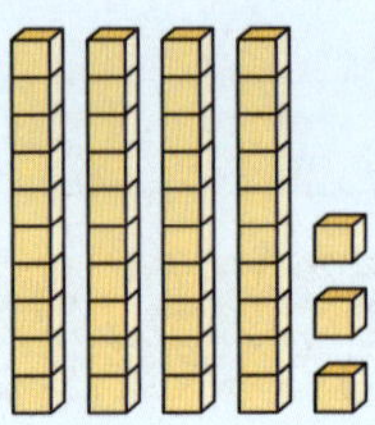

______ tens and ______ units

______ + ______ = ______

______

h 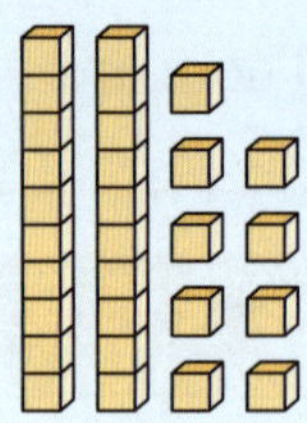

______ tens and ______ units

______ + ______ = ______

______

i 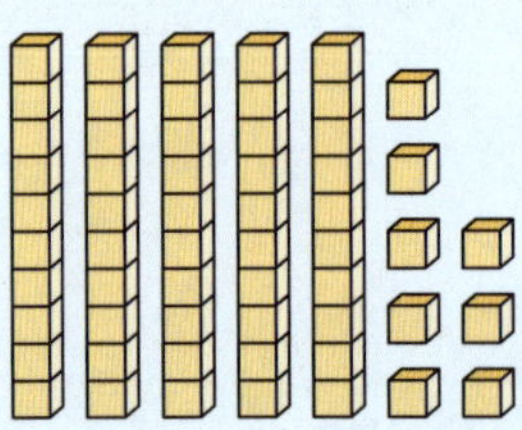

______ tens and ______ units

______ + ______ = ______

______

Exercise 18

Partition each number using place values:

3 8
3 tens 8 units
$30 + 8$
"thirty eight"

a 2 6
☐ tens ☐ units
☐ + ☐

b 1 9
☐ ten ☐ units
☐ + ☐

c 3 3
☐ tens ☐ units
☐ + ☐

d 4 7
☐ tens ☐ units
☐ + ☐

e 6 2
☐ tens ☐ units
☐ + ☐

f 9 5
☐ tens ☐ units
☐ + ☐

g 7 8
☐ tens ☐ units
☐ + ☐

h 5 4
☐ tens ☐ units
☐ + ☐

Activity

Complete this chart.

1	2		4	5		7		9	
11		13	14		16	17	18		20
	22	23		25	26		28	29	30
31	32		34	35	36	37		39	40
41		43	44		46	47	48	49	
51	52		54	55		57	58		60
	62	63	64		66	67		69	70
71	72	73		75	76		78	79	
81		83	84	85		87		89	90
	92	93	94		96	97	98		100

Exercise 19

Write these numbers in words:

a 25 ______________________

b 39 ______________________

c 46 ______________________

d 52 ______________________

e 69 ______________________

f 77 ______________________

g 91 ______________________

h 84 ______________________

Listening Activity

Click and follow the instructions.

1 ______ **2** ______

3 ______ **4** ______

5 ______ **6** ______

Discussion

A

______ apples

B

______ apples

Which bowl has more apples? ______

Which bowl has fewer apples? ______

$>$

5 is greater than 3.

5 is more than 3.

$<$

3 is less than 5.

3 is fewer than 5.

Exercise 20

Practise your writing.

> greater than

more than

< less than

fewer than

Exercise 21

Use *greater* or *less*.

a 

Six is ______________ than five.

b 

Four is ______________ than five.

c 

Eight is ______________ than nine.

d 

Seven is ______________ than five.

Exercise 22

Use *more* or *fewer*.

a 

Six is ______________ than four.

b 

Three is ______________ than five.

c 

Seven is ______________ than nine.

d 

Eight is ______________ than five.

Exercise 23

Use *greater* or *less*.

a $>$ means ______________ than.

b $<$ means ______________ than.

Exercise 24

Use $<$ or $>$.

a 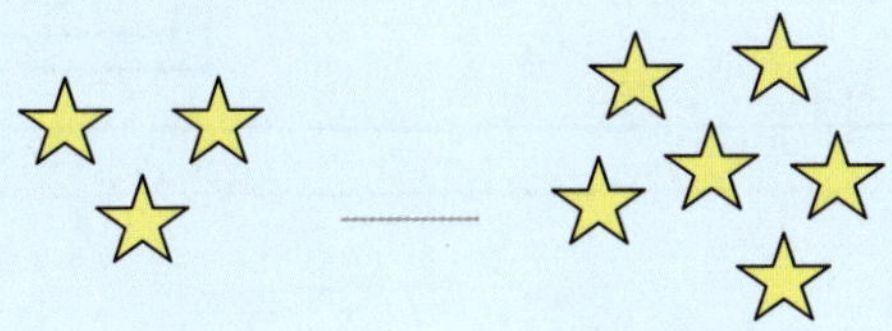

b 

c 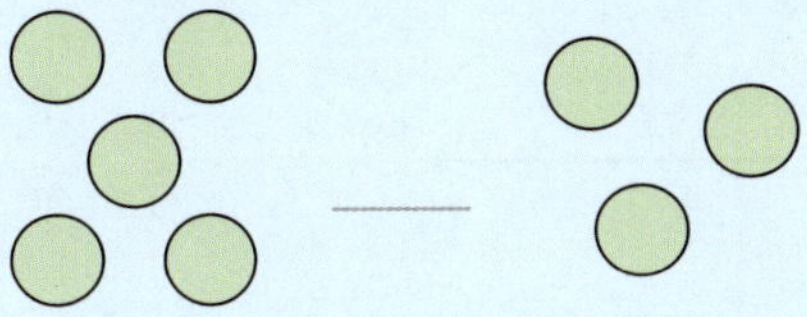

d 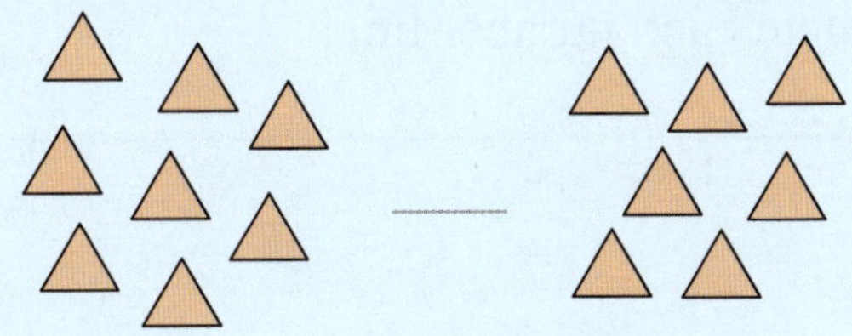

e 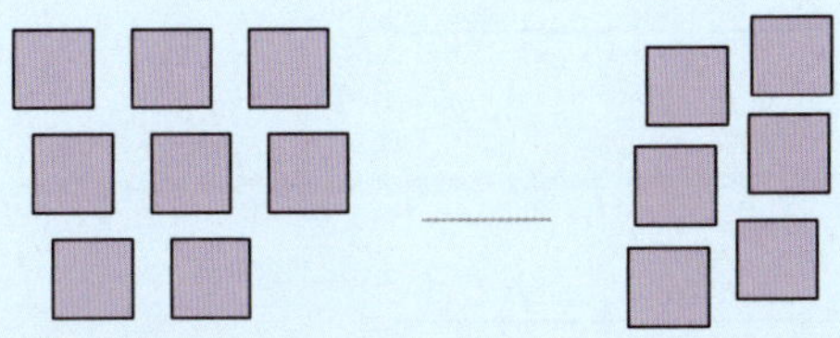

f 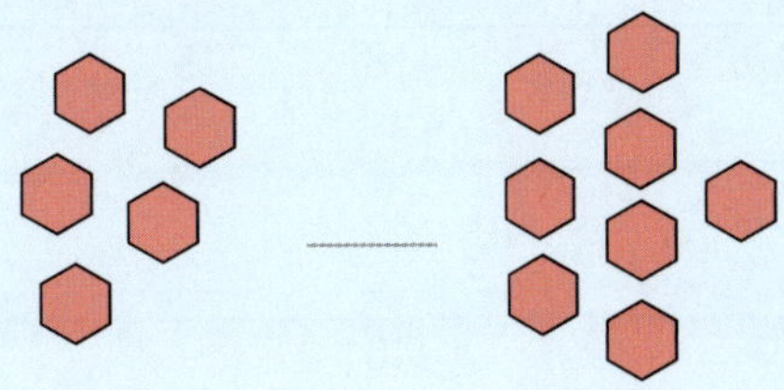

Exercise 25

Use $>$ or $<$.

a 5 ____ 7	**b** 1 ____ 7	**c** 9 ____ 4
d 3 ____ 2	**e** 2 ____ 5	**f** 6 ____ 2
g 4 ____ 10	**h** 8 ____ 6	**i** 7 ____ 9

Discussion

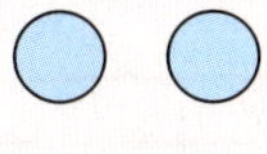

2

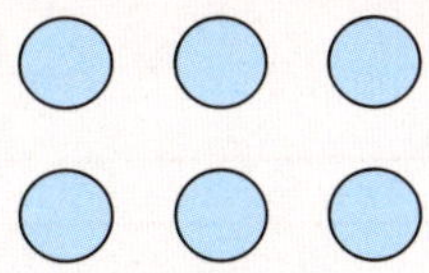

6

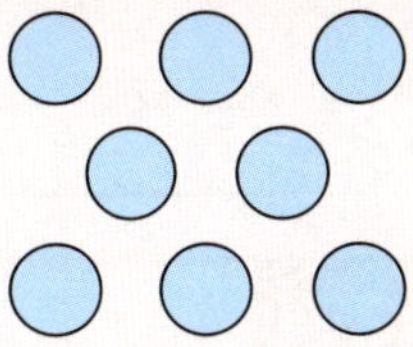

8

$2 < 6$ and $2 < 8$ so 2 is the *least* of the numbers.

$8 > 6$ and $8 > 2$ so 8 is the *greatest* of the numbers.

A **number line** lists whole numbers in order from least to greatest.

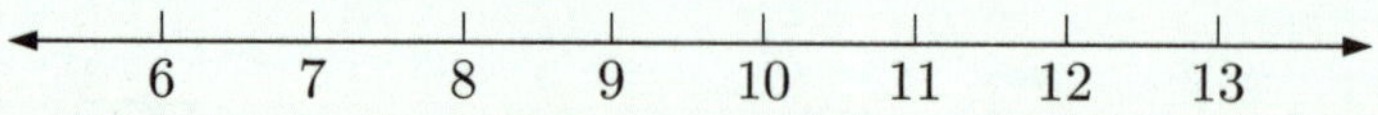

Exercise 26

Complete each number line:

a

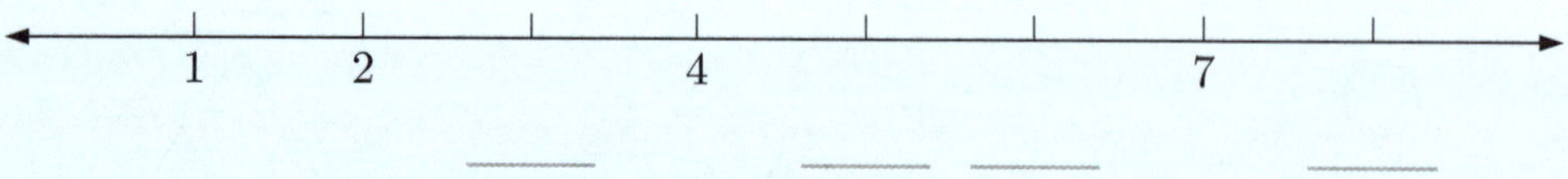

b

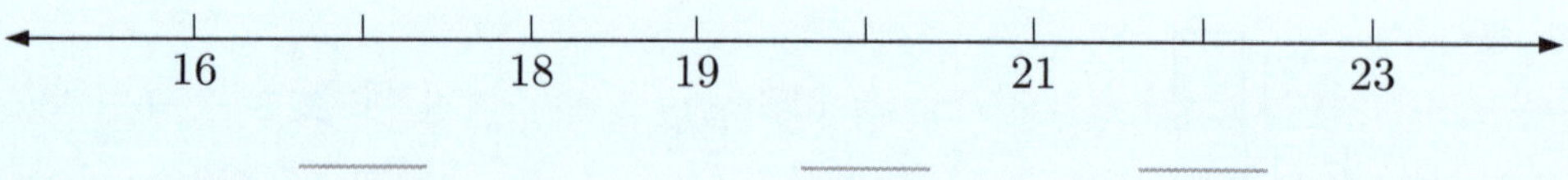

c

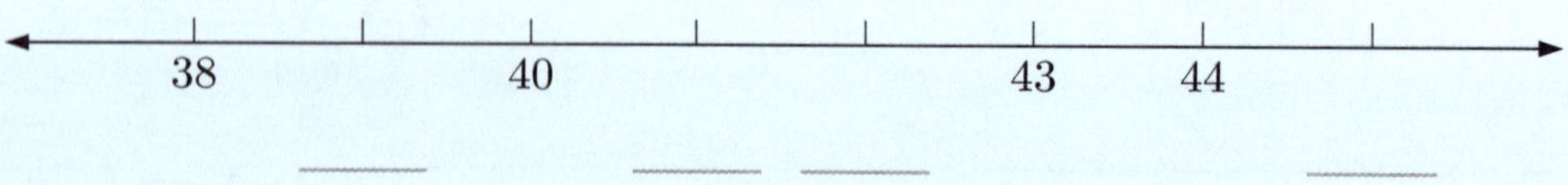

d

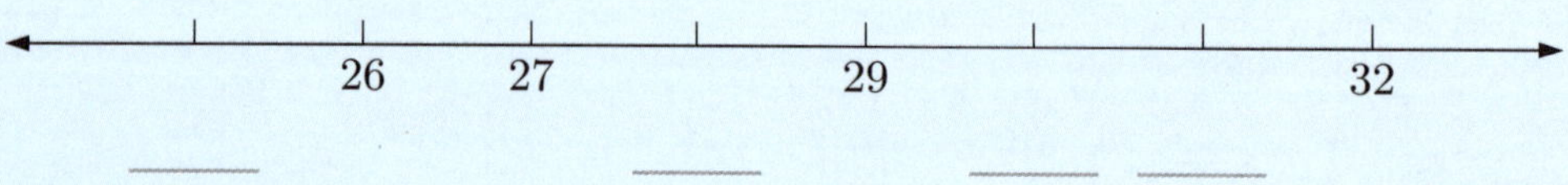

e

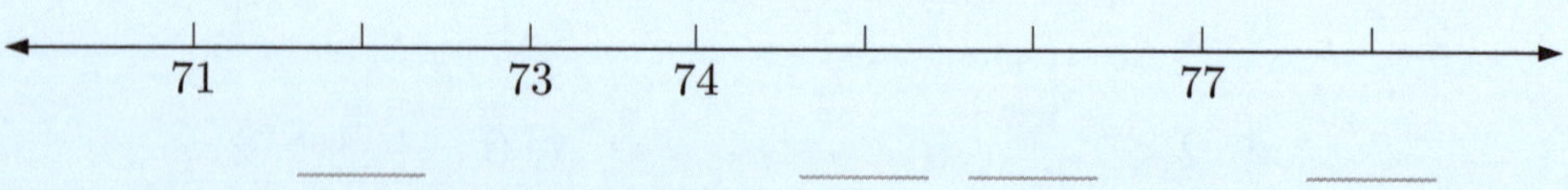

f

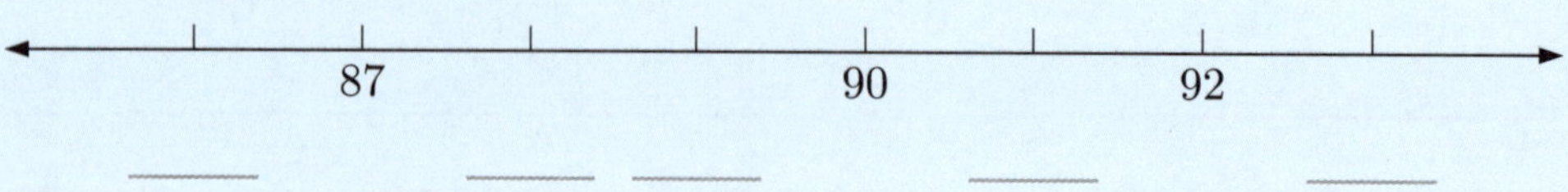

g

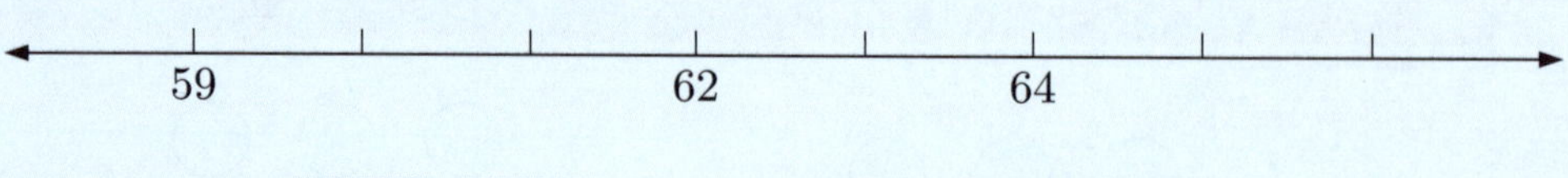

Exercise 27

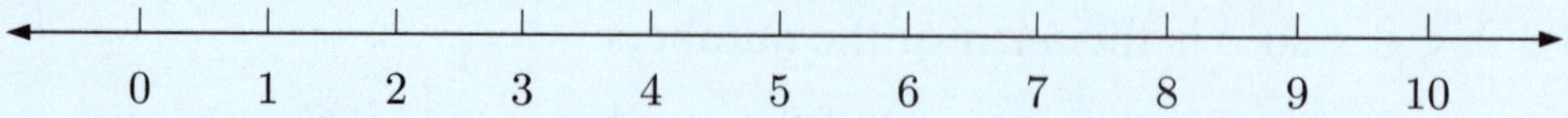

Circle the numbers *greater than* 5 in *red*.

Circle the numbers *less than* 5 in *blue*.

Exercise 28

a

1	2	3	4	5	6	7	8	9	10
11	12	13	14	15	16	17	18	19	20
21	22	23	24	25	26	27	28	29	30
31	32	33	34	35	36	37	38	39	40
41	42	43	44	45	46	47	48	49	50
51	52	53	54	55	56	57	58	59	60
61	62	63	64	65	66	67	68	69	70
71	72	73	74	75	76	77	78	79	80
81	82	83	84	85	86	87	88	89	90
91	92	93	94	95	96	97	98	99	100

Count up in 10s:

10, ________, ________, ________, ________, ________, ________,

________, ________, ________

b

1	2	3	4	5	6	7	8	9	10
11	12	13	14	15	16	17	18	19	20
21	22	23	24	25	26	27	28	29	30
31	32	33	34	35	36	37	38	39	40
41	42	43	44	45	46	47	48	49	50
51	52	53	54	55	56	57	58	59	60

Count up in 5s:

5, ________, ________, ________, ________, ________, ________,

________, ________, ________, ________, ________

c

1	2	3	4	5	6	7	8	9	10
11	12	13	14	15	16	17	18	19	20
21	22	23	24	25	26	27	28	29	30

Count up in 2s:

2, ________, ________, ________, ________, ________, ________,

________, ________, ________, ________, ________, ________,

________, ________

Exercise 29

Write these numbers in words:

a 55 ______________________________

b 43 ______________________________

c 98 ______________________________

d 37 ______________________________

e 86 ______________________________

f 62 ______________________________

g 100 ______________________________

Activity

Draw twenty three flowers.

Colour eleven of the flowers purple.

Exercise 30

What numbers are shown?

a

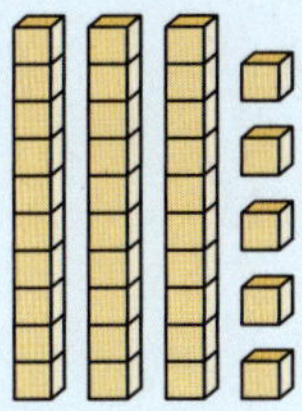

☐ tens and ☐ units

☐ + ☐ = ☐

b

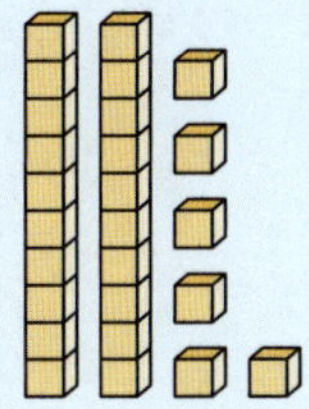

☐ tens and ☐ units

☐ + ☐ = ☐

c

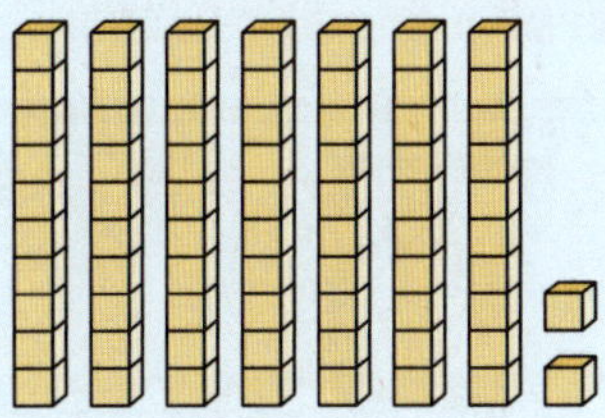

☐ tens and ☐ units

☐ + ☐ = ☐

Exercise 31

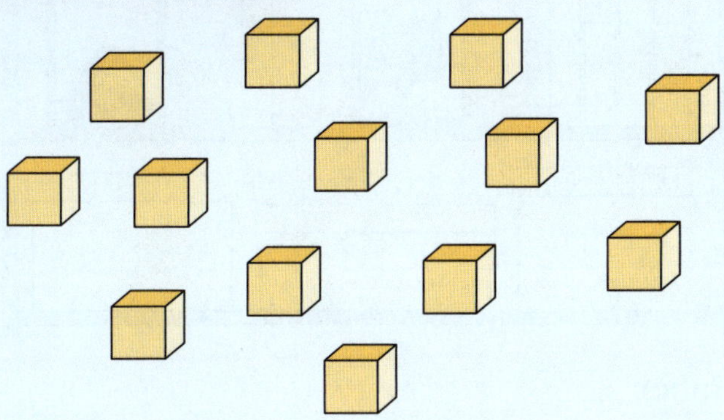

How many blocks are there?

Draw this number of blocks another way so they are easier to count.

Does your drawing make it easier to see the *place values*?

Activity

Some things I often count:

1 ....

2 ....

3 ....

Group Activity

a What objects can you count, that give the number ....

1 ....

2 ....

3 ....

4 ....

5 ....

6 ....

7 ....

8 ....

9 ....

10 ....

12 ....

18 ....

b What other numbers can you find?

....

....

CHAPTER 2: POSITION AND DIRECTION

We use many different words to describe *where* things are.

Exercise 1

The bird is to the *left* of the monkey.

left

____

____

The monkey is to the *right* of the bird.

right

____

____

Exercise 2

The kite is *above* Anna.

above

____

____

Anna is *below* the kite.

below

____

____

Exercise 3

The sun is *behind* the cloud. The cloud is *in front of* the sun.

behind in front of

Exercise 4

Complete each sentence using *left*, *right*, *above*, or *below*:

a The river is ______________ the bridge.

b The horse is to the ______________ of the river.

c The bird is ______________ the car.

d The car is to the ______________ of the river.

Discussion

What other words might we use to mean:

- "above"
- "below"?

Exercise 5

The bus is *alongside* the bus stop.

alongside

The bus is *next to* the bus stop.

next to

The bus stop is *near* to the store.

near

The school is *far* from the bus stop.

far

Exercise 6

Complete each sentence:

a The lemon is above the ____________.

b The strawberry is below the ____________.

c The peach is next to the ____________.

d The plum is alongside the ____________.

Discussion

What do we mean by an *opposite*?

Exercise 7

Match the words with *opposite* meanings.

far

left

behind

below

above

near

right

in front of

Exercise 8

Five people are waiting in a line.

Ben Sally Harry Erica Tim

Sally is in front of ______________.

Harry is behind ______________ and ______________.

Activity

What to do:

1 Write down the things you can see:

a to the left of the computer

....

b above the drawer.

....

2 Draw:

- a lamp to the right of the computer
- two pens next to the ruler
- a book in front of the computer
- a waste bin below the desk.

Exercise 9

3 people I sit *near to* in class are

________________, ________________, and ________________.

3 people I sit *far from* in class are

________________, ________________, and ________________.

Activity

Complete this car park scene by drawing:

- a light above the grey car
- a blue car to the right of the green car
- a lady in front of the red car
- a mouse under the grey car
- a man next to the green car.

Exercise 10

A lift goes *up* and *down*.

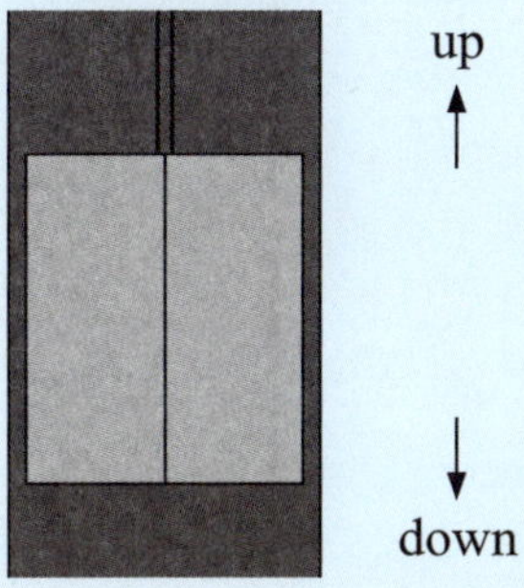

up down

Discussion

What other words do we use to describe *directions*?

Exercise 11

Which way should the pointer be moved to click on the:

a letter

b recycle bin

c photo

d internet browser?

Exercise 12

The sandwich is *inside* the lunch box.

inside

The mango is *outside* the lunch box.

outside

Discussion

List some games or sports you might play inside, outside, or either.

Inside	*Outside*	*Either*

CHAPTER 3: ADDITION

$$3 + 4 = 7$$

"three plus four equals seven"

Exercise 1

These words tell us to **add**. Practise your writing.

add	sum	plus	total
add	sum	plus	total
________	________	________	________
________	________	________	________

What is $2 + 3$?

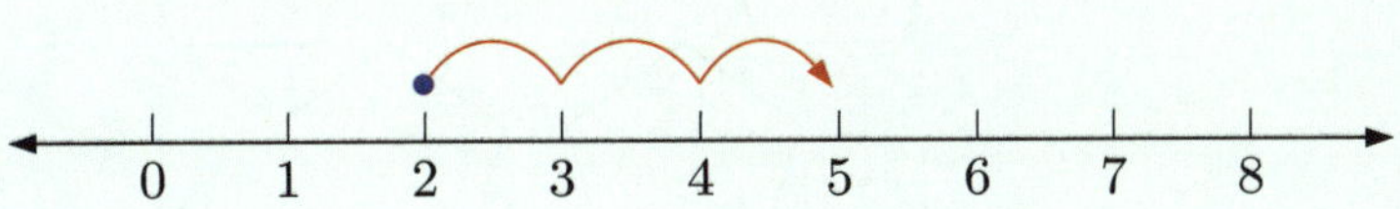

$2 + 3 = 5$

Say together:

"Two add three equals five."

"Two plus three equals five."

"The sum of two and three is five."

"The total of two and three is five."

Exercise 2

Use the number line to help you add:

a $4+2=$ ______

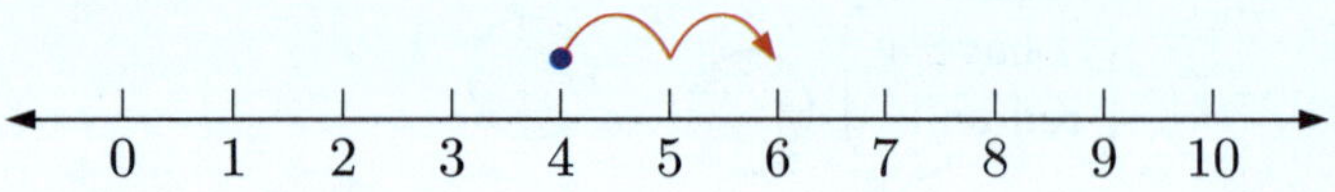

b $1+3=$ ______

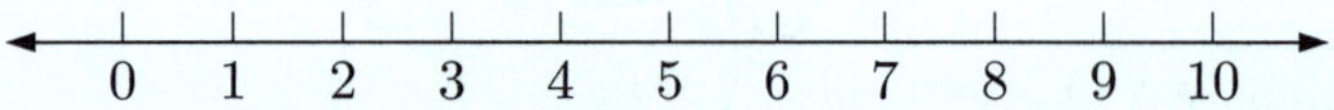

c $2+4=$ ______

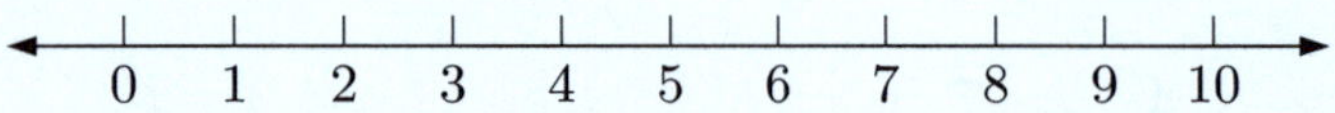

d $3+5=$ ______

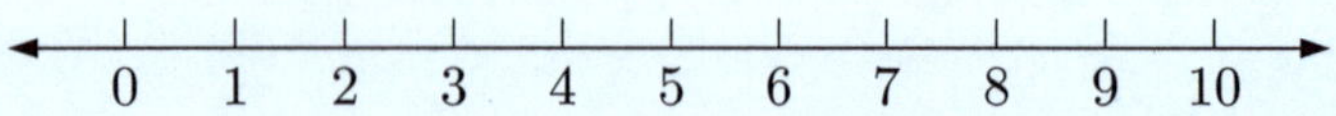

e $5+2=$ ______

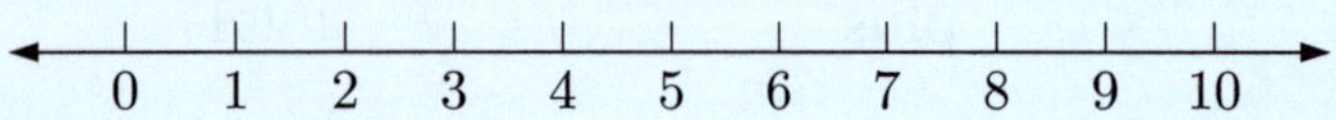

Exercise 3

Colour the objects to match the sum.

a $4+1=5$

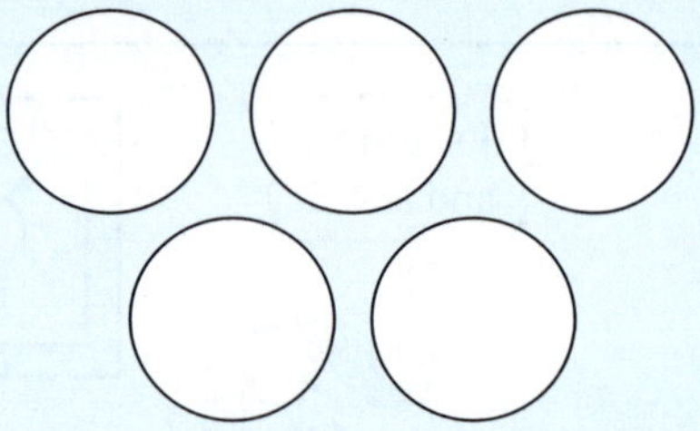

b $2+3=5$

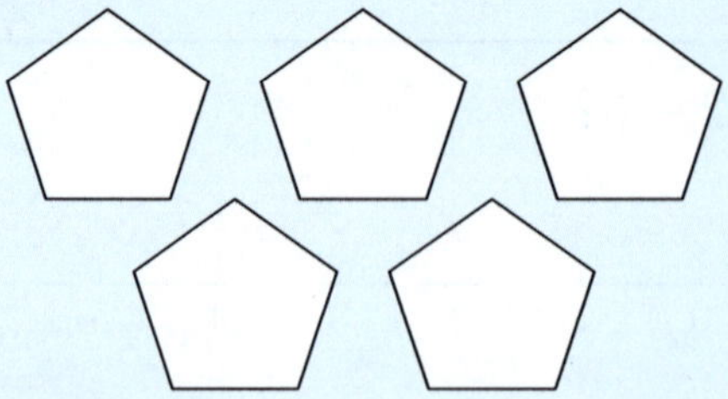

c $3+3=6$

d $5+3=8$

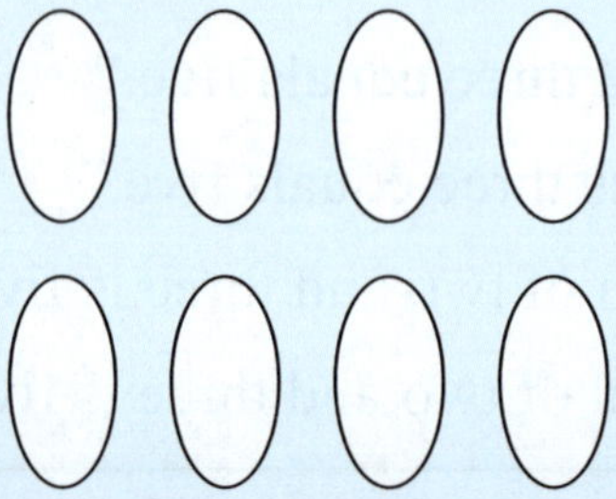

Exercise 4

Find the sum:

a

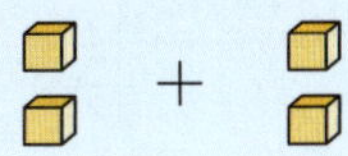

$2+2=$ ______

b

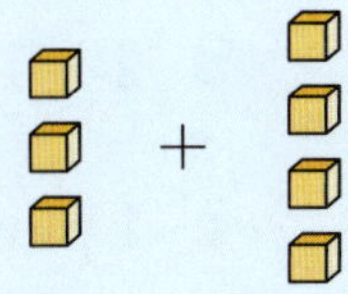

$3+4=$ ______

c

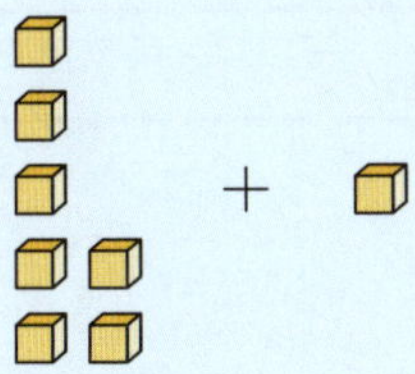

$7+1=$ ______

d

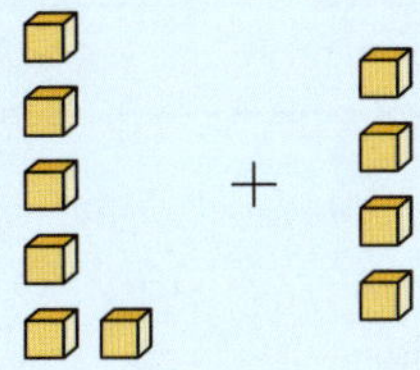

$6+4=$ ______

Exercise 5

Draw and colour objects to match the sum.

a $3+2=5$

b $4+3=7$

c $2+5=7$

d $5+4=9$

e $3+6=9$

f $7+2=9$

Exercise 6

Find the sum:

a $3+2=$ ______ **b** $2+1=$ ______ **c** $2+5=$ ______

d 4 add $2=$ ______ **e** 1 plus $5=$ ______ **f** 2 add $2=$ ______

g $4+4=$ ______ **h** 6 plus $2=$ ______ **i** $4+6=$ ______

Exercise 7

a The sum of 3 and 3 is ______.

b The total of 2 and 4 is ______.

c The sum of 6 and 3 is ______.

d The total of 4 and 5 is ______.

e The sum of 8 and 2 is ______.

f The total of 2 and 7 is ______.

g The sum of 7 and 1 is ______.

h The total of 5 and 5 is ______.

Exercise 8

Write pairs of numbers that add to 4.

$4=\square+\square$ $4=\square+\square$

$4=\square+\square$

Exercise 9

Write pairs of numbers that total 7.

$7=\square+\square$ $7=\square+\square$ $7=\square+\square$

$7=\square+\square$ $7=\square+\square$ $7=\square+\square$

Exercise 10

Write pairs of numbers that total 10.

10 = ☐ + ☐   10 = ☐ + ☐   10 = ☐ + ☐

10 = ☐ + ☐   10 = ☐ + ☐   10 = ☐ + ☐

10 = ☐ + ☐   10 = ☐ + ☐   10 = ☐ + ☐

Exercise 11

Use the number line to help you add:

a 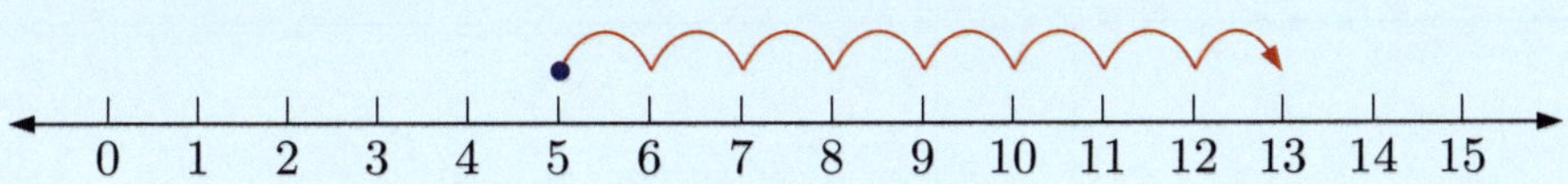

$5 + 8 =$ ______

b 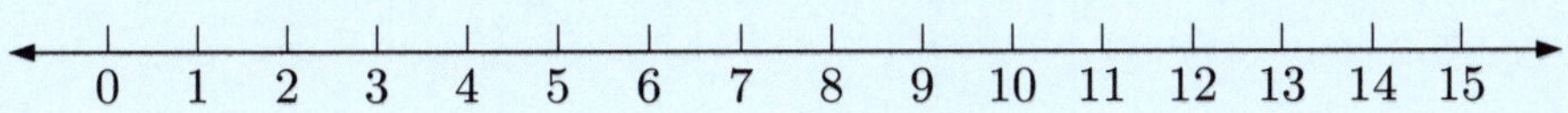

$7 + 5 =$ ______

c 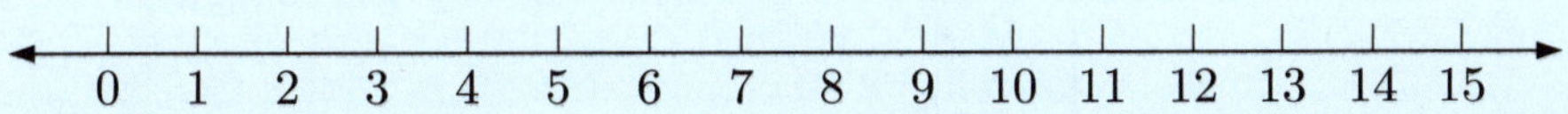

$8 + 4 =$ ______

d 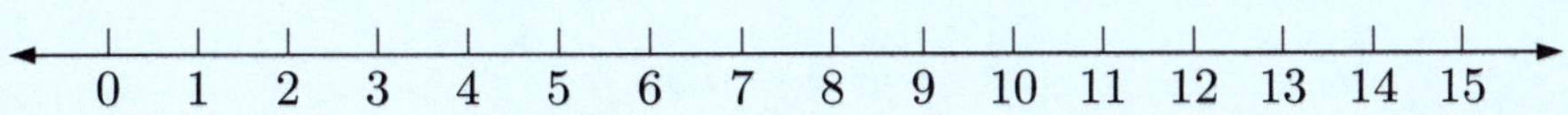

$9 + 5 =$ ______

e 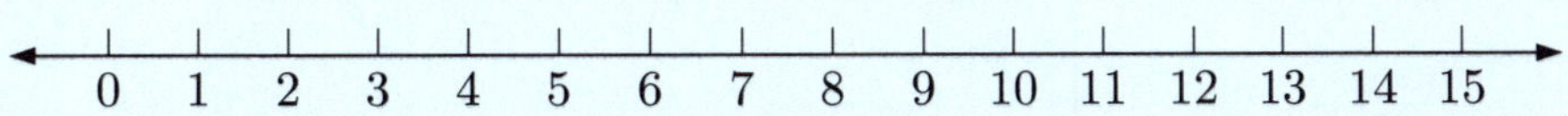

$5 + 7 =$ ______

f 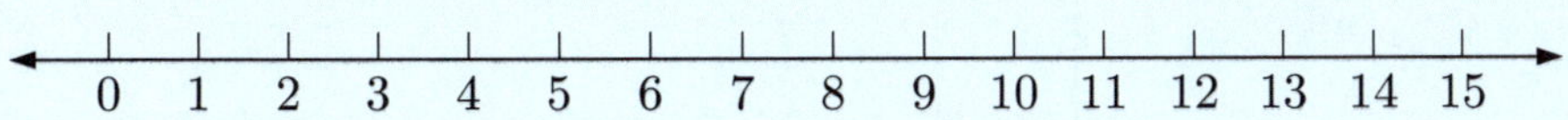

$4 + 9 =$ ______

g 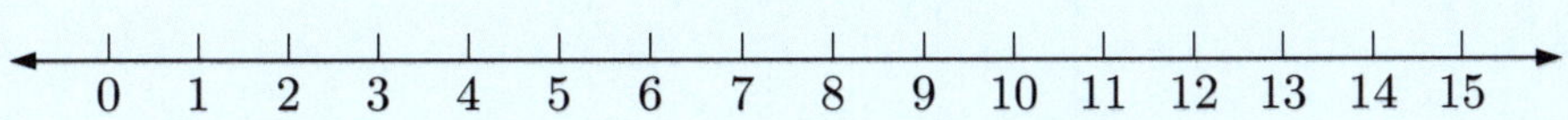

$6 + 6 =$ ______

Exercise 12

Colour the objects to match the sum.

a $7 + 6 = 13$

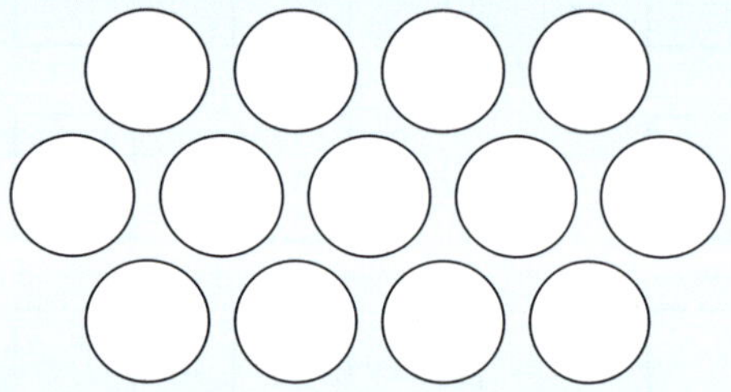

b $8 + 7 = 15$

Exercise 13

Draw and colour objects to match the sum.

a $5 + 7 = 12$

b $7 + 4 = 11$

c $9 + 9 = 18$

d $7 + 9 = 16$

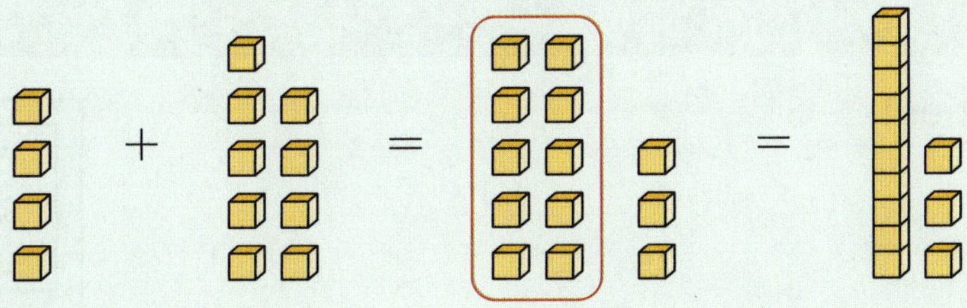

$4 + 9 = 10 + 3 = 13$

Exercise 14

Complete:

a

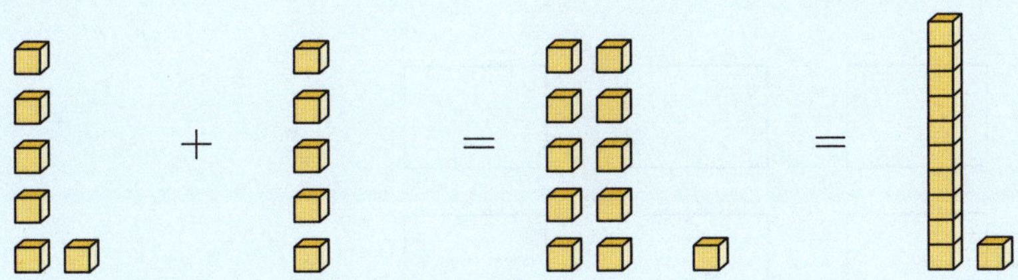

$\_\_\_\_ + \_\_\_\_ = 10 + \_\_\_\_ = \_\_\_\_$

b

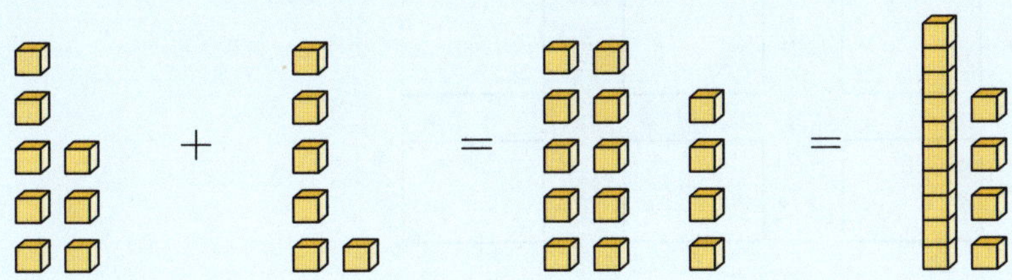

$\_\_\_\_ + \_\_\_\_ = 10 + \_\_\_\_ = \_\_\_\_$

c

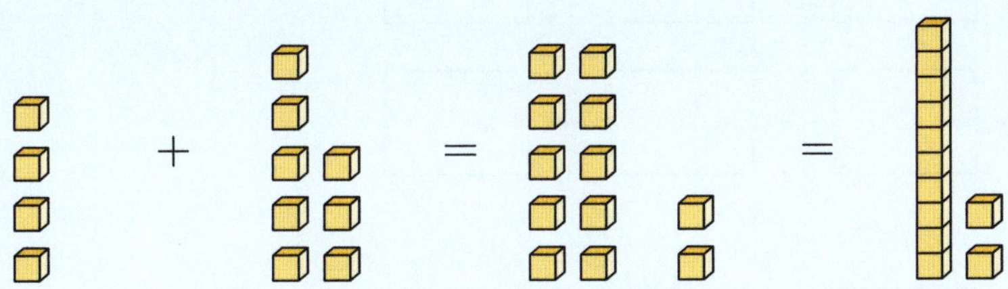

$\_\_\_\_ + \_\_\_\_ = 10 + \_\_\_\_ = \_\_\_\_$

Exercise 15

Use blocks to help you add:

a $7 + 6 =$ ______ **b** $8 + 5 =$ ______ **c** $5 + 6 =$ ______

d $6 + 8 =$ ______ **e** $3 + 9 =$ ______ **f** $7 + 7 =$ ______

Game

Click the icon to practise your addition!

Exercise 16

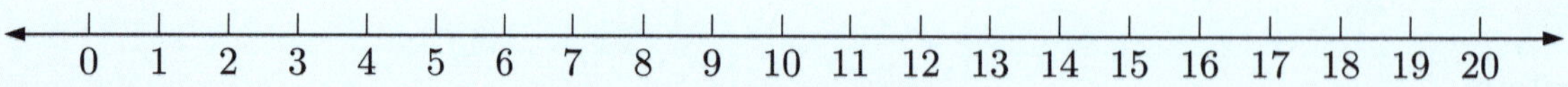

Add 1 each time:

a 3 → 4 → ☐ → ☐ → ☐

b 8 → ☐ → ☐ → ☐ → ☐

Add 2 each time:

c 2 → 4 → ☐ → ☐ → ☐

d 5 → ☐ → ☐ → ☐ → ☐

Add 3 each time:

e 2 → ☐ → ☐ → ☐ → ☐

f 6 → ☐ → ☐ → ☐ → ☐

Add 4 each time:

g 2 → ☐ → ☐ → ☐ → ☐

h 3 → ☐ → ☐ → ☐ → ☐

Add 5 each time:

i 2 → ☐ → ☐ → ☐

Add 6 each time:

j 1 → ☐ → ☐ → ☐

Exercise 17

Describe each sequence.

a 3 → 5 → 7 → 9 → 11

I start with ______, and add ______ each time.

b 1 → 5 → 9 → 13 → 17

I start with ______, and add ______ each time.

c 3 → 6 → 9 → 12 → 15

____

Exercise 18

These dice show a "double 4".

"Double 4" is $4 + 4 = 8$.

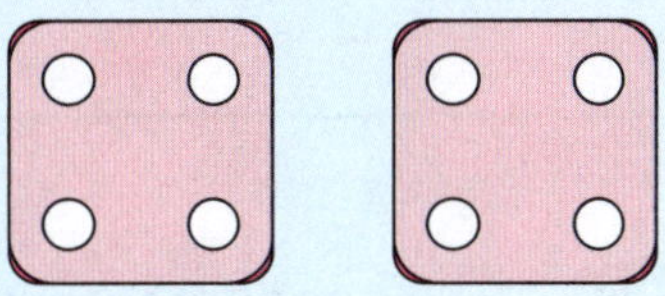

Complete this table of doubles:

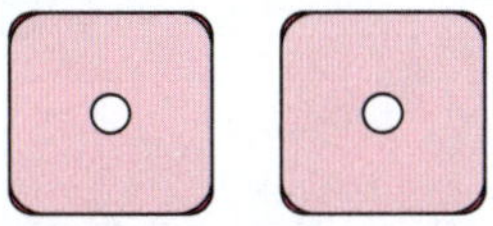	double 1	$1 + 1 =$ ______
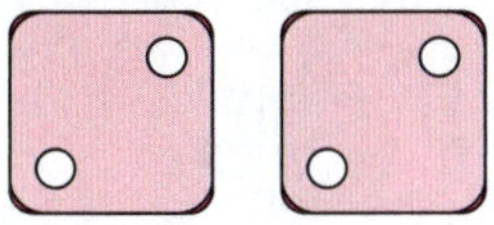		
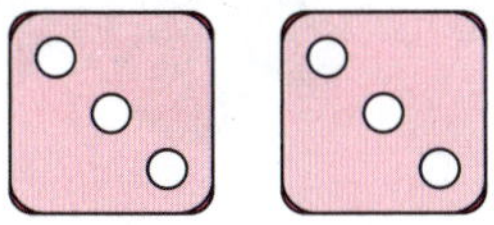		
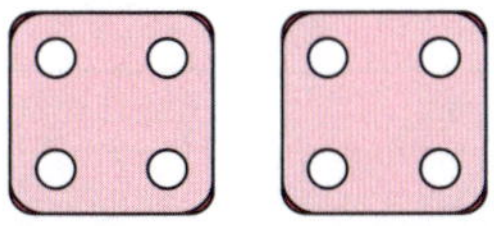		
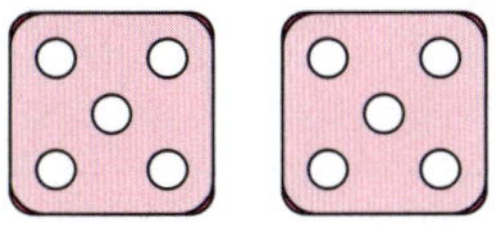		
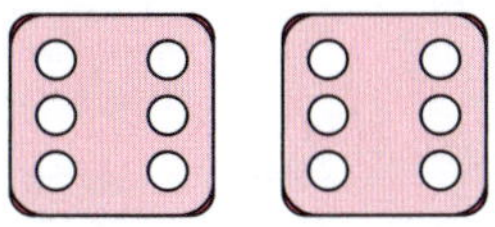		

Exercise 19

Complete these pairs of additions.

a
$3 + 2 =$ ______
$2 + 3 =$ ______

b
$4 + 5 =$ ______
$5 + 4 =$ ______

c
$2 + 8 =$ ______
$8 + 2 =$ ______

d
$5 + 7 =$ ______
$7 + 5 =$ ______

Discussion

Can you see patterns in the pairs of additions above?

Can you explain *why* this happens?

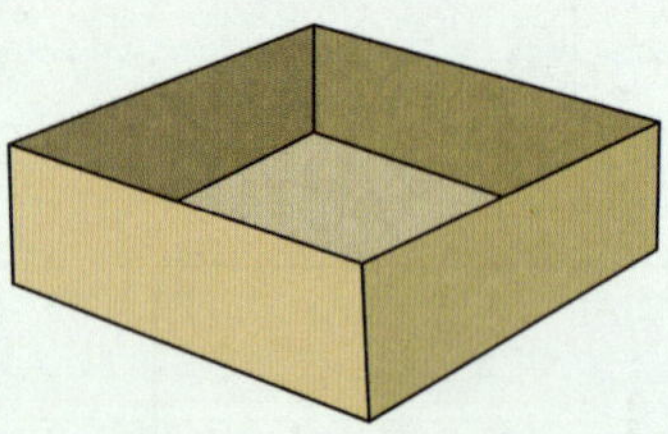

There are no elephants in the box.

We say there are **zero** elephants or 0 elephants.

$0 + 3 = 3$

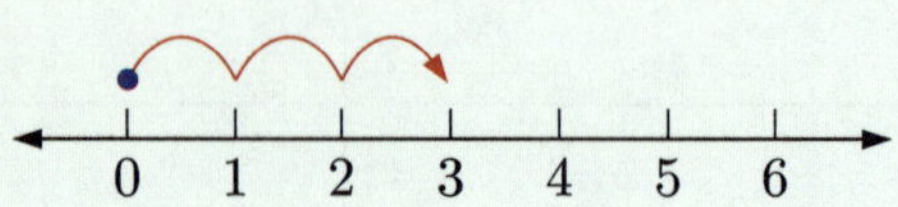

I start with nothing and add 3.

$3 + 0 = 3$

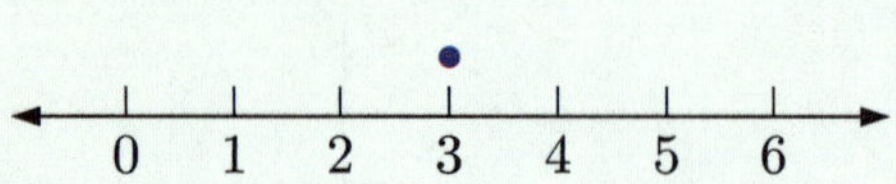

I start with 3 and add nothing.

Exercise 20

Complete these pairs of additions.

a $2 + 0 =$ ______
$0 + 2 =$ ______

b $0 + 6 =$ ______
$6 + 0 =$ ______

c $9 + 0 =$ ______
$0 + 9 =$ ______

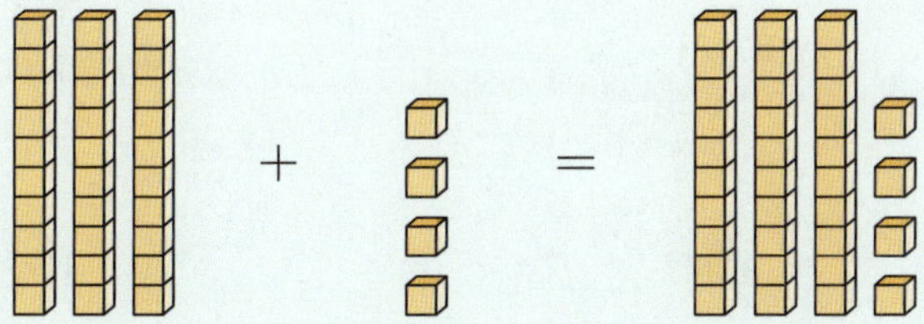

$30 \quad + \quad 4 \quad = \quad 34$

Exercise 21

Complete these additions.

$0+2=$ ______

$10+2=$ ______

$20+2=$ ______

$30+2=$ ______

$40+2=$ ______

$0+5=$ ______

$10+5=$ ______

$20+5=$ ______

$30+5=$ ______

$40+5=$ ______

$0+7=$ ______

$20+7=$ ______

$40+7=$ ______

$60+7=$ ______

$80+7=$ ______

Exercise 22

Complete these additions.

$3+0=$ ______

$3+10=$ ______

$3+20=$ ______

$3+30=$ ______

$1+0=$ ______

$1+20=$ ______

$1+40=$ ______

$1+60=$ ______

$9+10=$ ______

$9+20=$ ______

$9+40=$ ______

$9+60=$ ______

Exercise 23

Complete these additions.

a $3 + \_\_\_\_ = 10$ **b** $5 + \_\_\_\_ = 10$ **c** $8 + \_\_\_\_ = 10$

d $6 + \_\_\_\_ = 10$ **e** $\_\_\_\_ + 2 = 10$ **f** $\_\_\_\_ + 9 = 10$

g $\_\_\_\_ + 4 = 10$ **h** $1 + \_\_\_\_ = 10$ **i** $\_\_\_\_ + 7 = 10$

Exercise 24

Complete these additions.

$3 + 6 = \_\_\_\_$	$6 + 4 = \_\_\_\_$	$8 + 6 = \_\_\_\_$
$3 + 7 = \_\_\_\_$	$6 + 5 = \_\_\_\_$	$8 + 7 = \_\_\_\_$
$3 + 8 = \_\_\_\_$	$6 + 6 = \_\_\_\_$	$8 + 8 = \_\_\_\_$
$3 + 9 = \_\_\_\_$	$6 + 7 = \_\_\_\_$	$8 + 9 = \_\_\_\_$
$3 + 10 = \_\_\_\_$	$6 + 8 = \_\_\_\_$	$8 + 10 = \_\_\_\_$

Exercise 25

Complete these additions.

a $5 + \_\_\_\_ = 7$ **b** $\_\_\_\_ + 3 = 6$ **c** $2 + \_\_\_\_ = 5$

d $4 + \_\_\_\_ = 9$ **e** $\_\_\_\_ + 1 = 8$ **f** $9 + \_\_\_\_ = 11$

g $8 + \_\_\_\_ = 13$ **h** $\_\_\_\_ + 7 = 12$ **i** $6 + \_\_\_\_ = 14$

Exercise 26

$4 + 2 + 5 = 11$

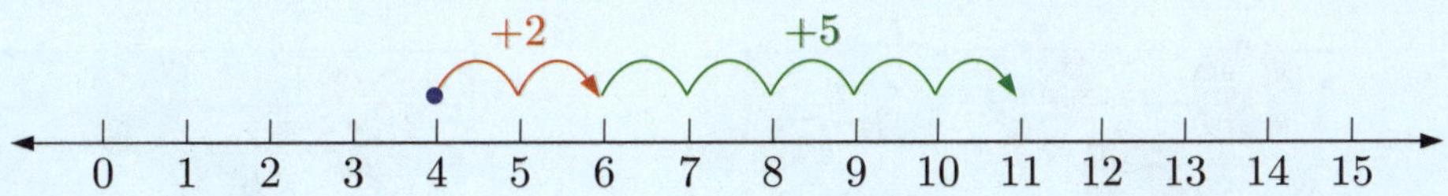

Complete these additions.

a $1+2+3=$ ______ **b** $2+1+2=$ ______ **c** $1+3+3=$ ______

d $2+4+1=$ ______ **e** $3+3+3=$ ______ **f** $4+2+5=$ ______

g $6+1+4=$ ______ **h** $4+7+3=$ ______ **i** $8+2+3=$ ______

Game

Click the icon to practise your addition.

Exercise 27

Amber has 3 black horses and 2 grey horses.

In total, Amber has $3+2=5$ horses.

Write an addition to solve these problems.

a Li has 4 blue shirts and 3 white shirts.

In total, Li has ____________ = ______ shirts.

b Pat has 5 bowls and 4 plates.

In total, Pat has ____________ = ______ items.

c Sally has 8 hair clips and 4 ribbons.

In total, Sally has ____________ = ______ items.

d Terry has 6 trees and 9 shrubs.

Altogether, Terry has ____________ = ______ plants.

Exercise 28

Write an addition to find the total number of:

a green items ....

b red items ....

c apples ....

d bell peppers ....

Exercise 29

$2 + 3 = 5$

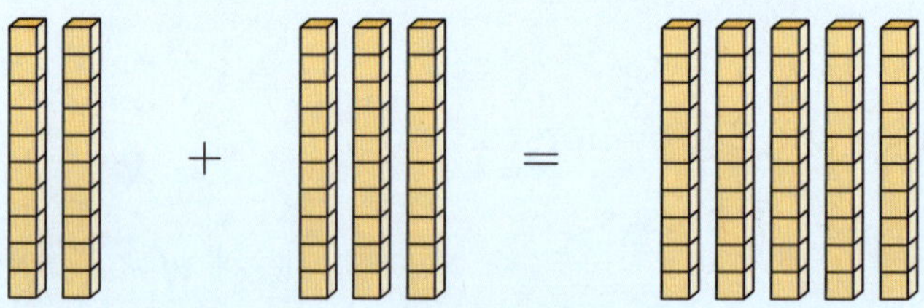

$20 + 30 = 50$

Complete these additions.

a $1 + 4 =$ ....
$10 + 40 =$ ....

b $5 + 2 =$ ....
$50 + 20 =$ ....

c $4 + 3 =$ ....
$40 + 30 =$ ....

d $3 + 6 =$ ....
$30 + 60 =$ ....

e $6 + 2 =$ ....
$60 + 20 =$ ....

f $8 + 1 =$ ....
$80 + 10 =$ ....

g $5 + 4 =$ ....
$50 + 40 =$ ....

h $4 + 6 =$ ....
$40 + 60 =$ ....

i $7 + 2 =$ ....
$70 + 20 =$ ....

CHAPTER 4: SHAPE

A **straight** line never changes direction.

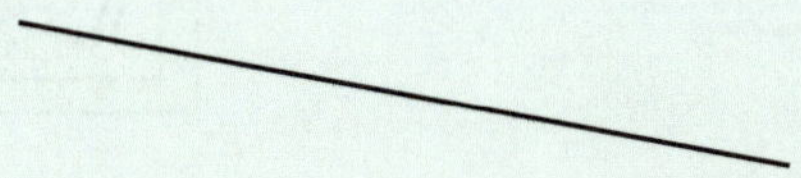

A **curve** is not straight.

Discussion

Look around your classroom.

What objects can you see that have:

- straight lines
- curves?

Exercise 1

Practise your writing.

straight

straight

____

curve

curve

____

Exercise 2

Colour in *green* the shapes with straight sides.

Colour in *blue* the shapes with curves.

A shape is **2-dimensional** if it can be drawn on a **surface**.

Exercise 3

These 2-dimensional shapes have straight sides.

Practise your writing.

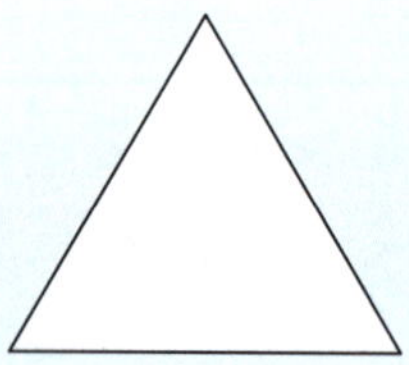

triangle

triangle ______________

A triangle has _____ straight sides.

square

square ______________

A square has _____ straight sides.

rectangle

rectangle ______________

A rectangle has _____ straight sides.

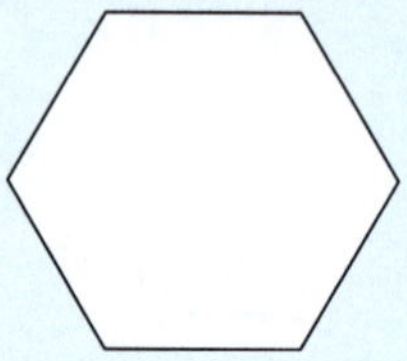

hexagon

hexagon ______________

A hexagon has _____ straight sides.

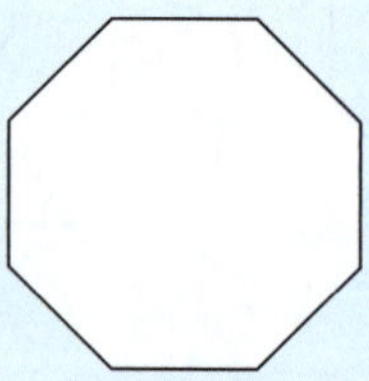

octagon

octagon ______________

An octagon has _____ straight sides.

Each side of a shape can also be called an **edge**.

Two edges meet at a **corner**.

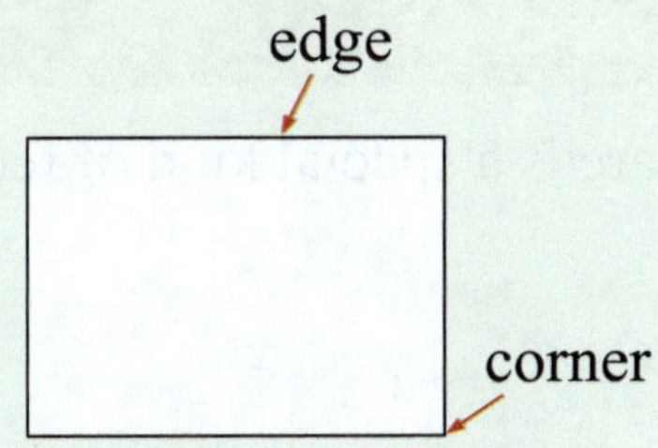

Exercise 4

Practise your writing.

edge

edge

____

corner

corner

____

Exercise 5

Complete each sentence using *more* or *fewer*:

A rectangle has ____________ edges than a triangle.

A square has ____________ edges than a hexagon.

A hexagon has ____________ edges than a triangle.

Exercise 6

These shapes are made using a curve. Practise your writing.

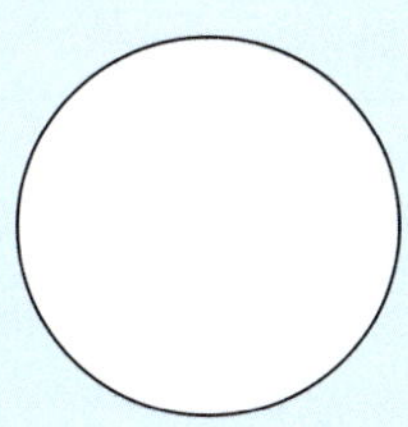

circle

circle

____

____

oval

oval

____

____

Discussion

A square is a special kind of rectangle.

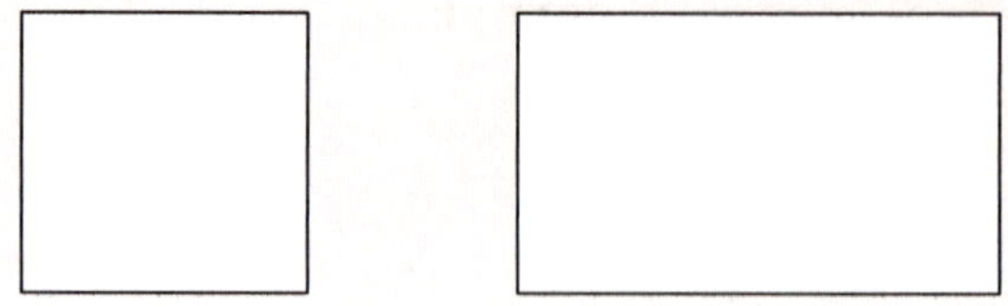

square rectangle

What is *special* about the square?

....

Exercise 7

Count the shapes. There are .... squares and .... triangles.

Discussion

An octagon has .... sides.

An octopus has .... legs.

Exercise 8

Colour the hexagons blue and the octagons orange.

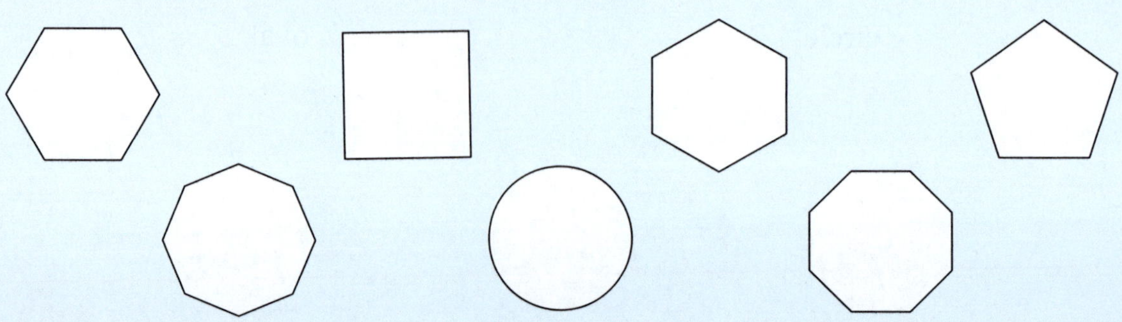

Exercise 9

Draw each shape. Use your ruler if needed.

a a square

b a circle

c a triangle

d an oval

e a hexagon

f a rectangle

Exercise 10

Describe the shape of the sign.

a

b

c

Exercise 11

Complete:

Shape	*Edges*	*Corners*
rectangle	4	
triangle		
		6
square		
	8	

Exercise 12

a I have 3 sides.
I have 3 corners.
My sides are all straight.

I am a ______________________.

b I am drawn with 1 curve.
I look like a dinner plate.

I am a ______________________.

c I have 6 sides.
I have 6 corners.
My sides are all straight.

I am a ______________________.

Discussion

How would you describe a *square* to someone?

I have ______ sides.

My sides have the same ____________________ .

I have ______ corners.

The corners look the same.

Listening Activity

You will need a pencil, a ruler, and a piece of scrap paper.

Listen to the audio. Follow the instructions.

Activity

This pattern is made of squares and triangles.

Colour the shapes with different colours.

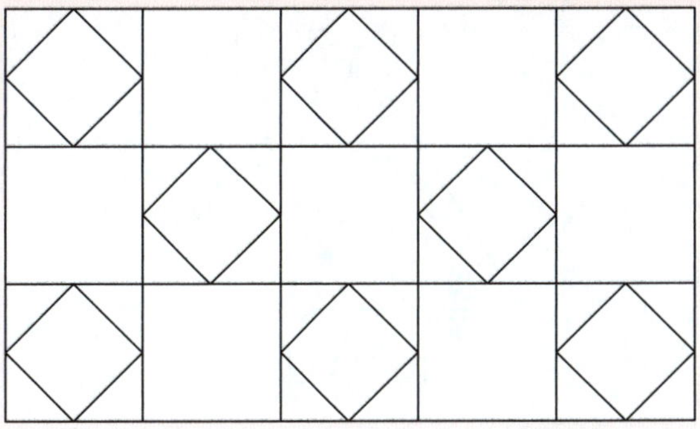

Draw a pattern using squares and rectangles.

Colour it with different colours.

A **solid** is a shape which takes up **space**.

We say that solids are **3-dimensional**.

We can *draw* solids as 2-dimensional shapes on a page.

Exercise 13

Here are some 3-dimensional solids. Practise your writing.

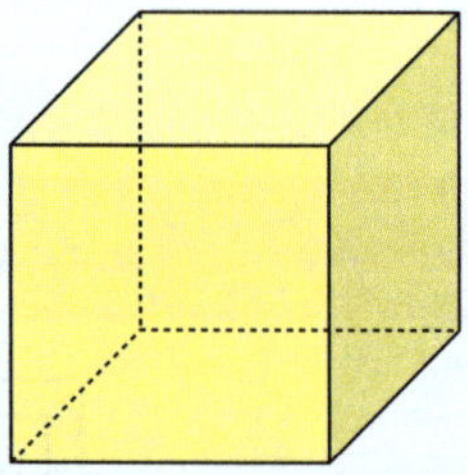

cube

cube

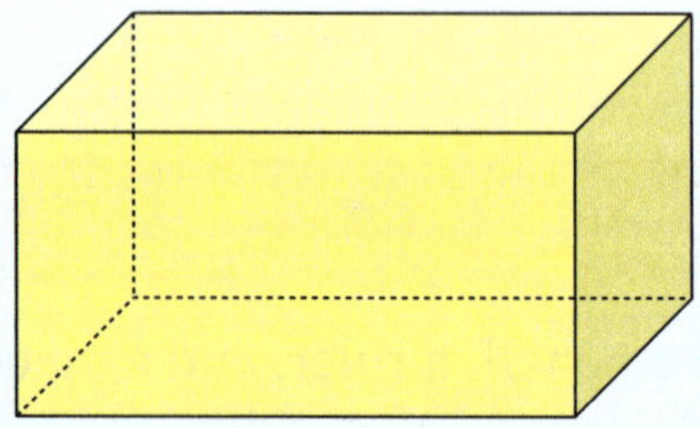

rectangular prism

rectangular prism

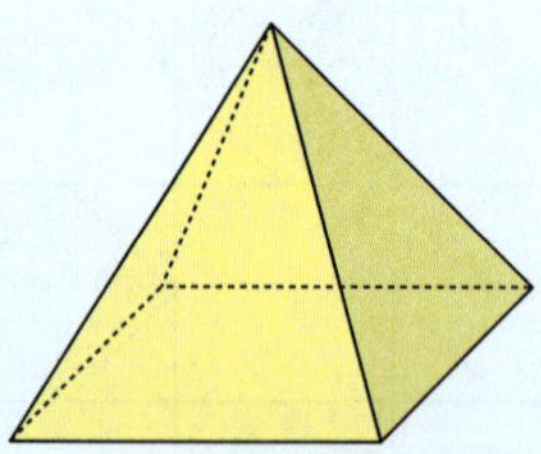

pyramid

pyramid

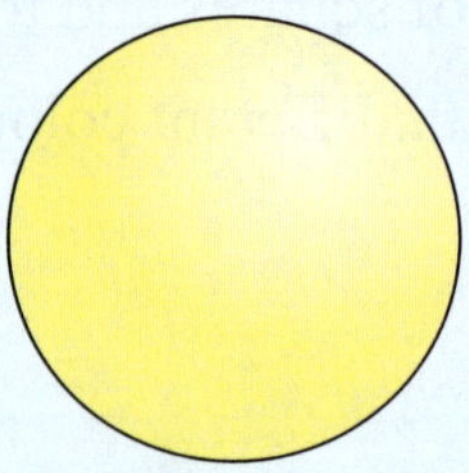

sphere

sphere

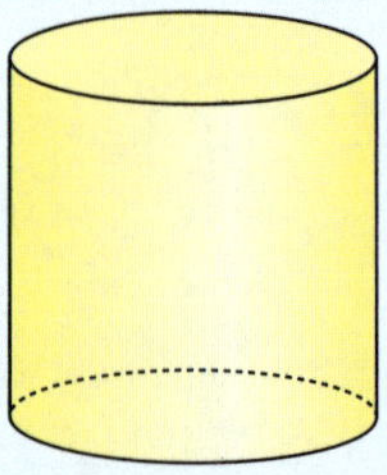

cylinder

cylinder

Exercise 14

Name the solid you see in the picture.

a

b

c

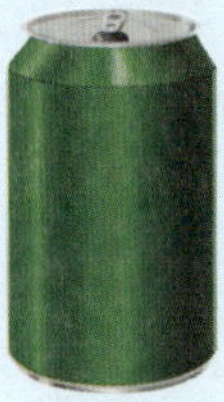

d

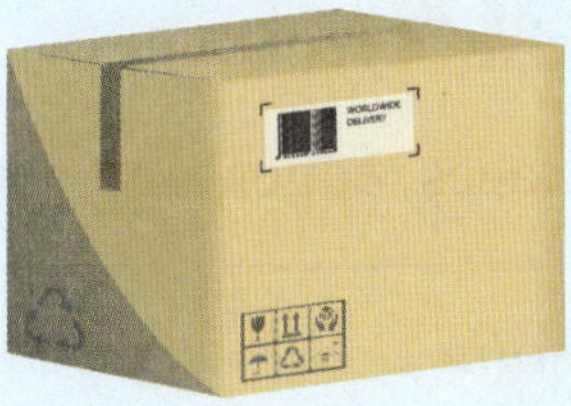

e

Discussion

Look around the classroom.

Record examples of solids that you see in your classroom.

Any flat surface of a solid is called a ____________ .

Any line where two faces meet is called an ____________ .

Any point where three faces meet is called a ____________ .

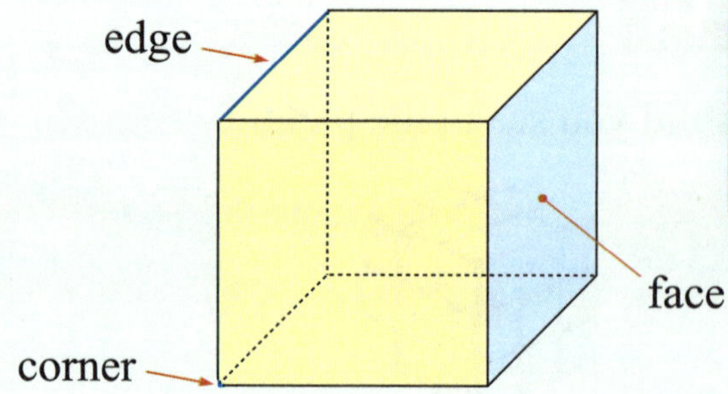

Exercise 15

Count the faces, edges, and corners of each solid:

Solid	*Faces*	*Edges*	*Corners*
cube			
rectangular prism			
pyramid			
sphere	1 curved surface		
cylinder	______ faces and 1 curved surface		

Puzzle

How many triangles can you find?

______

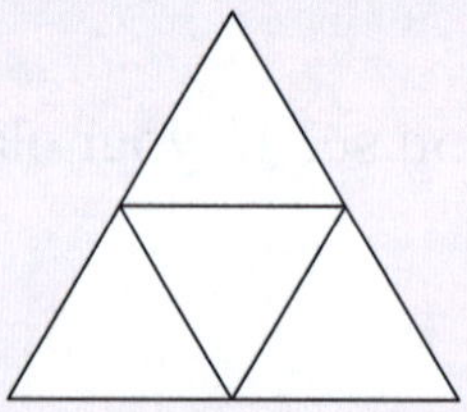

How many squares can you find?

______

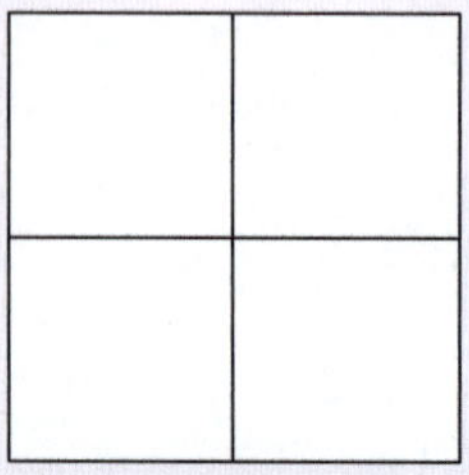

Exercise 16

Complete the sentences.

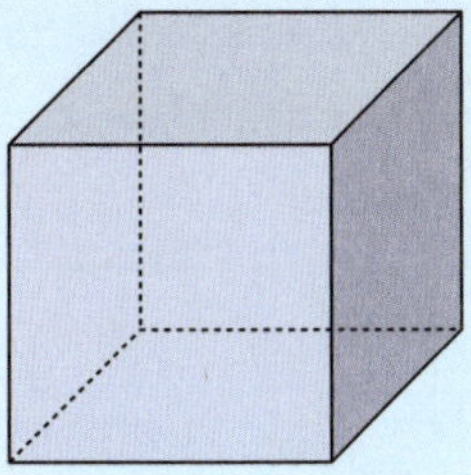

The faces of a cube are all

________________________________ .

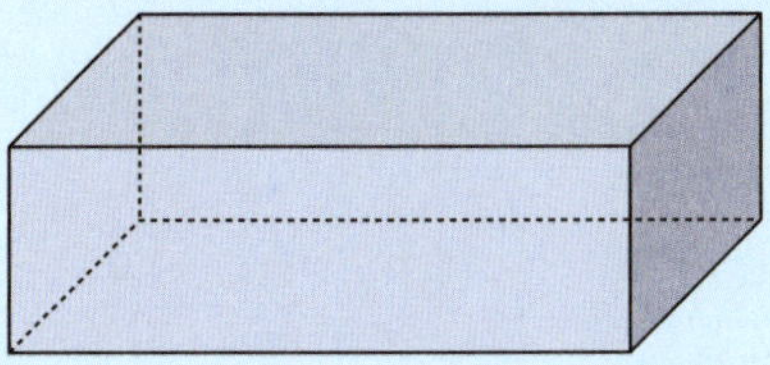

The faces of a rectangular prism are all

________________________________ .

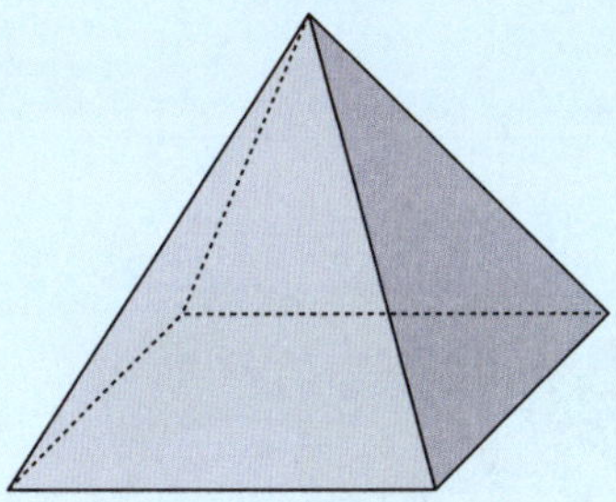

The base of this pyramid is a

________________________________ .

The other faces of the pyramid are

________________________________ .

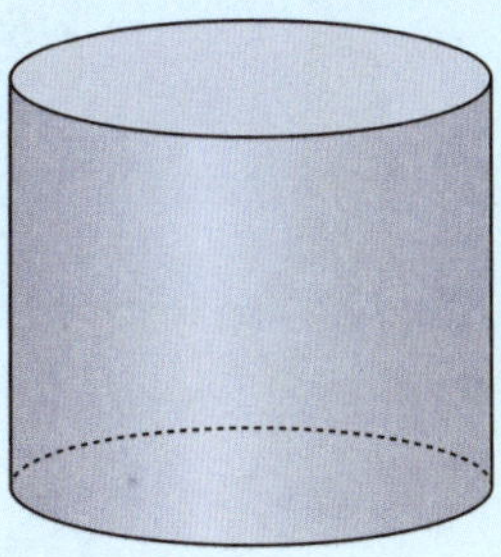

The two flat surfaces of a cylinder are

________________________________ .

Activity

1 Use popsicle sticks and plasticine to make your own:

a cube **b** rectangular prism **c** pyramid.

2 Make a sphere using plasticine.

Activity

In the picture below, describe where you see:

a a triangle ....

b a square ....

c a rectangle ....

d a circle ....

e a rectangular prism ....

f a sphere ....

g a pyramid ....

h a cylinder ....

CHAPTER 5: SUBTRACTION

I have 6 eggs.	I cook 4 eggs for lunch.	I have 2 eggs left.

$$6 - 4 = 2$$

"six minus four equals two"

Exercise 1

These words tell us to **subtract**. Practise your writing.

subtract	minus	take away
subtract	minus	take away
______________	______________	______________

Exercise 2

Use the pictures to write a subtraction:

a

_____ − _____ = _____

b

_____ − _____ = _____

c

_____ − _____ = _____

d

_____ − _____ = _____

What is $7 - 4$?

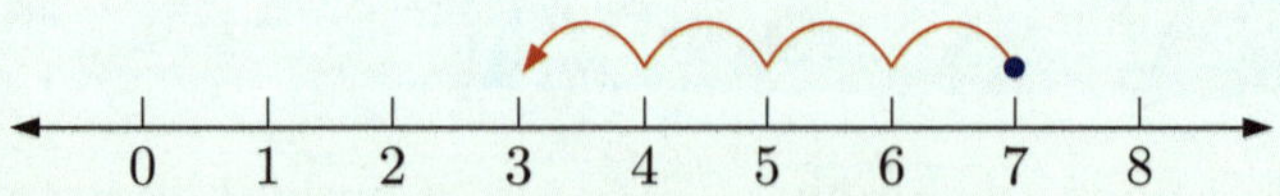

$7 - 4 = 3$

Say together:

"Seven subtract four equals three."

"Seven minus four equals three."

"Seven take away four equals three."

Exercise 3

Count backwards on the number line to help you subtract:

a $5 - 3 =$ ______

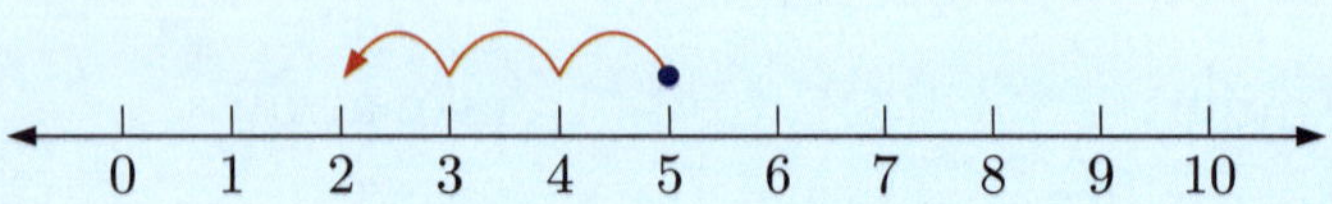

b $8 - 2 =$ ______

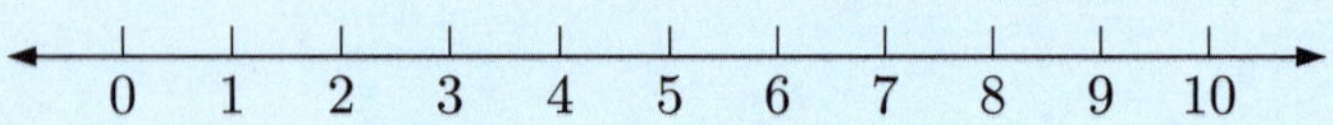

c $7 - 5 =$ ______

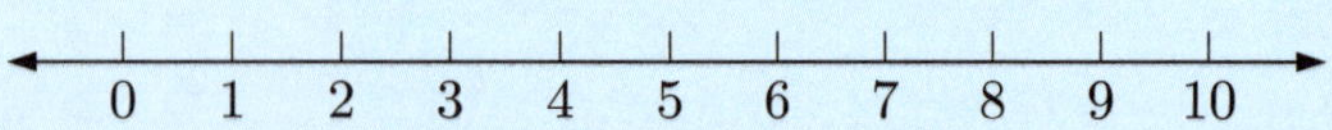

d $9 - 6 =$ ______

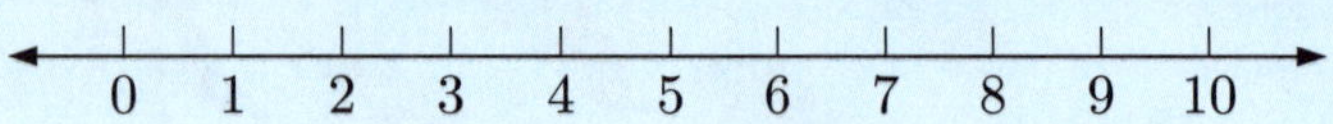

e $6 - 4 =$ ______

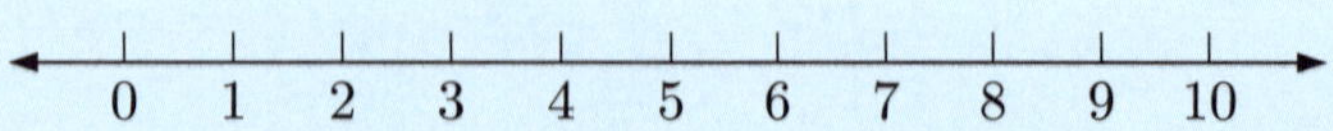

f $10 - 5 =$ ______

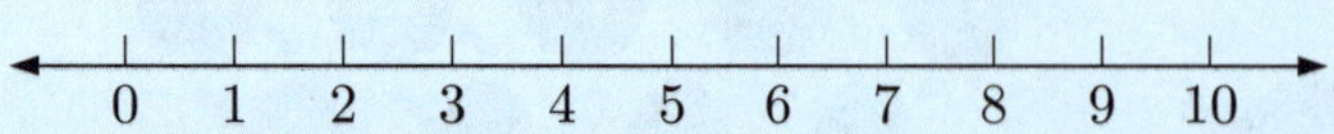

g $8 - 7 =$ ______

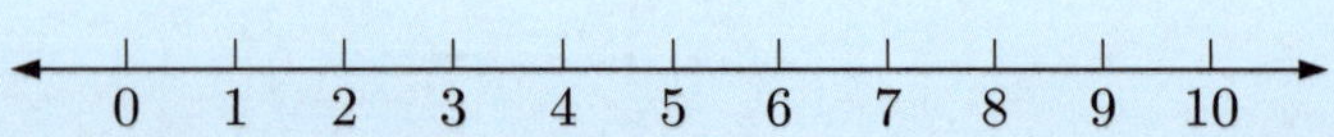

Exercise 4

Cross out pictures to help you subtract.

a

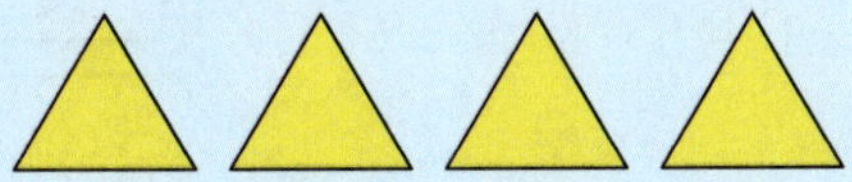

$4 - 2 =$ ______

b

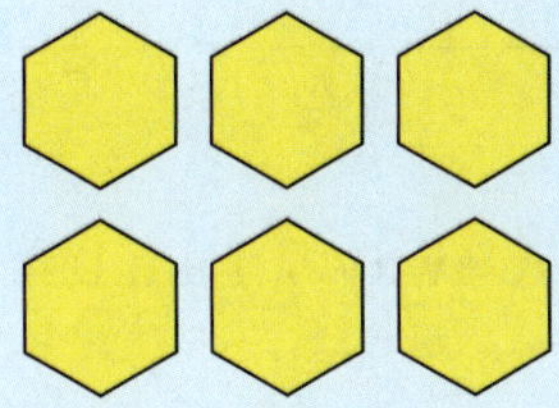

$6 - 1 =$ ______

c

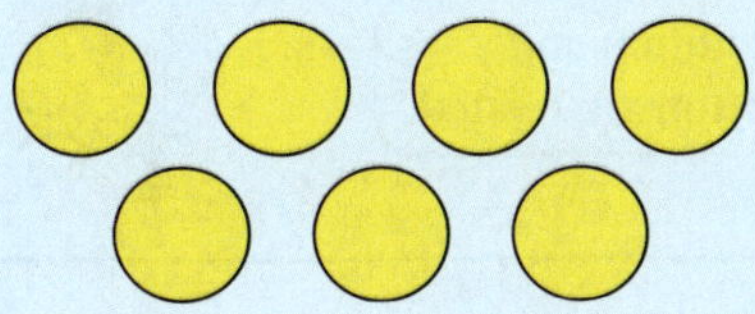

$7 - 3 =$ ______

d

$10 - 6 =$ ______

e

$8 - 5 =$ ______

f

$9 - 2 =$ ______

Exercise 5

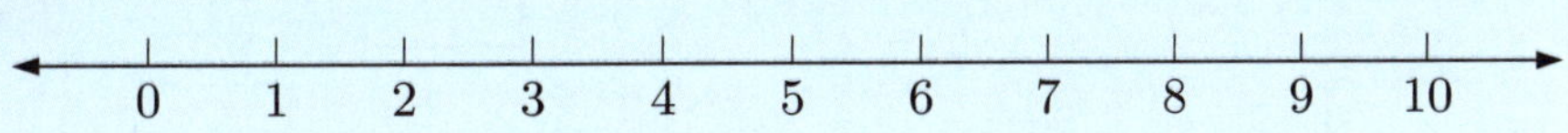

Use the number line to help you subtract.

a $3 - 1 =$ ______ **b** $4 - 2 =$ ______ **c** $5 - 4 =$ ______

d $7 - 4 =$ ______ **e** $4 - 1 =$ ______ **f** $5 - 2 =$ ______

g $6 - 3 =$ ______ **h** $8 - 3 =$ ______ **i** $7 - 1 =$ ______

j $9 - 7 =$ ______ **k** $9 - 3 =$ ______ **l** $8 - 4 =$ ______

m $6 - 4 =$ ______ **n** $7 - 6 =$ ______ **o** $8 - 5 =$ ______

To get to 6 from 1, I need to add 5.

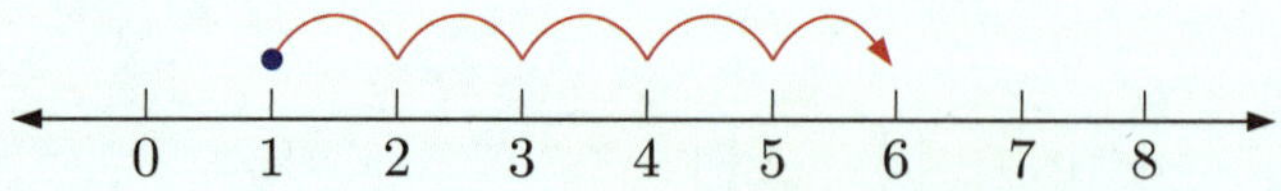

$6 = 1 + 5$

If I start with 6 and take away 5, I am left with 1.

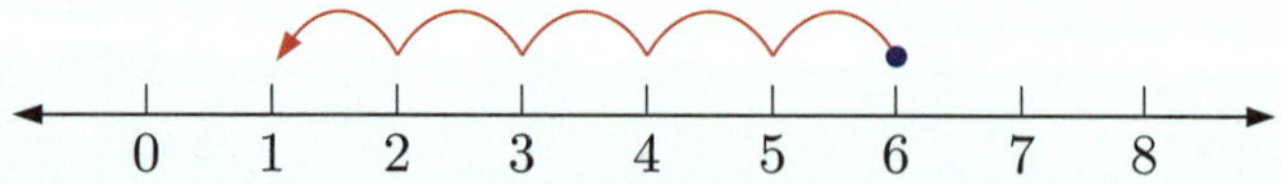

$6 - 5 = 1$

Exercise 6

Find the missing numbers.

a $6 = \square + 2$

so $6 - 2 = \square$.

b $4 = \square + 3$

so $4 - 3 = \square$.

c $7 = \square + 2$

so $7 - 2 = \square$.

d $9 = \square + 1$

so $9 - 1 = \square$.

e $8 = \square + 5$

so $8 - 5 = \square$.

f $10 = \square + 4$

so $10 - 4 = \square$.

Exercise 7

Complete the subtraction.

a 6 take away 4 = ______

b 7 minus 5 = ______

c 5 subtract 1 = ______

d 9 take 6 = ______

e 7 minus 3 = ______

f 8 subtract 7 = ______

Discussion

"Subtract 5 from 8" means ______ – ______

"Take 2 away from 7" means ______ – ______

Exercise 8

Complete the subtraction.

a Subtract 3 from 5. ______ – ______ = ______

b Take 7 away from 9. ______ – ______ = ______

c Subtract 3 from 6. ______ – ______ = ______

d Take 2 away from 8. ______ – ______ = ______

Exercise 9

Write a subtraction to solve each problem:

a I have 5 apples. I eat 2 apples.

______ – ______ = ______

I have ______ apples left.

b Sophie has 7 pens. 4 pens are blue.

______ – ______ = ______

______ pens are not blue.

c A choir has 8 singers. 2 singers are away sick.

______ – ______ = ______

______ singers are well.

d Otieno has 9 goats. He gives 3 goats to his neighbour.

______ – ______ = ______

Otieno has ______ goats left.

e I have 6 blueberries. I feed 2 to my lizard.

______ – ______ = ______

I have ______ blueberries left.

Exercise 10

$2+4=6$
so $6-2=4$
and $6-4=2$

Use the addition to write *two* subtractions.

a $3+2=5$

so $5-$ ______ $=$ ______

and $5-$ ______ $=$ ______

b $3+5=8$

so $8-$ ______ $=$ ______

and $8-$ ______ $=$ ______

c $4+5=$ ______

so ______ $-4=$ ______

and ______ $-5=$ ______

d $6+3=$ ______

so ______ $-6=$ ______

and ______ $-3=$ ______

Anna was given 5 lollies.
She has eaten none.
She has 5 lollies left.

$5-0=5$

Ben was given 5 lollies.
He has eaten them all.
He has no lollies left.

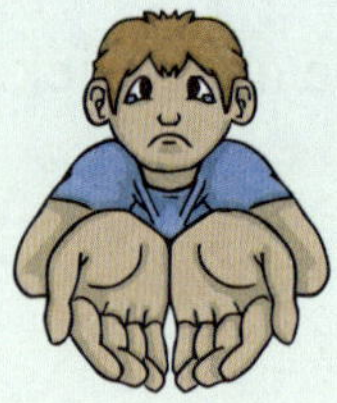

$5-5=0$

Exercise 11

Complete these subtractions.

a $3-0=$ ______ **b** $2-2=$ ______ **c** $4-0=$ ______

d $6-6=$ ______ **e** $9-0=$ ______ **f** $7-7=$ ______

g $8-0=$ ______ **h** $4-4=$ ______

Exercise 12

Complete these subtractions.

a

$3 - 3 =$ ....

$4 - 3 =$ ....

$5 - 3 =$ ....

$6 - 3 =$ ....

$7 - 3 =$ ....

$8 - 3 =$ ....

$9 - 3 =$ ....

$10 - 3 =$ ....

b

$5 - 5 =$ ....

$6 - 5 =$ ....

$7 - 5 =$ ....

$8 - 5 =$ ....

$9 - 5 =$ ....

$10 - 5 =$ ....

$11 - 5 =$ ....

$12 - 5 =$ ....

Exercise 13

Use the number line to help with these subtractions.

$10 - 0 =$ ....

$10 - 1 =$ ....

$10 - 2 =$ ....

$10 - 3 =$ ....

$10 - 4 =$ ....

$10 - 5 =$ ....

$10 - 6 =$ ....

$10 - 7 =$ ....

$10 - 8 =$ ....

$10 - 9 =$ ....

$10 - 10 =$ ....

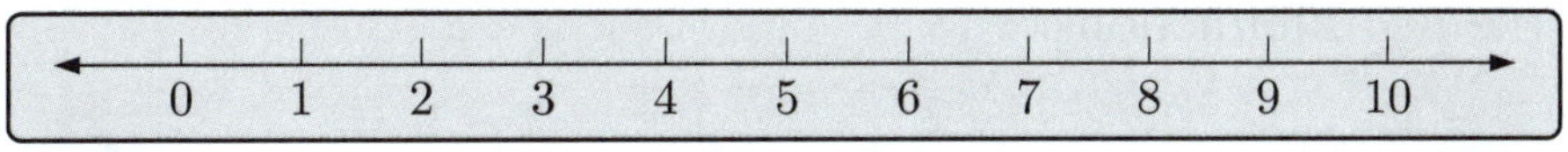

Exercise 14

Use the number line to help you subtract:

a 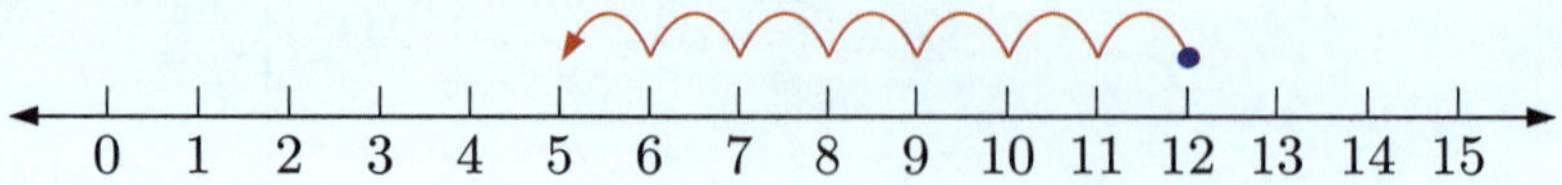

$12 - 7 =$ ______

b 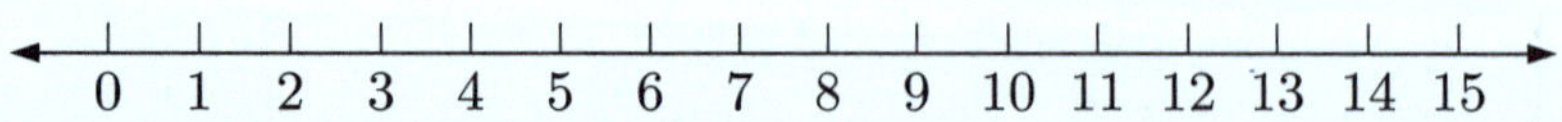

$11 - 4 =$ ______

c 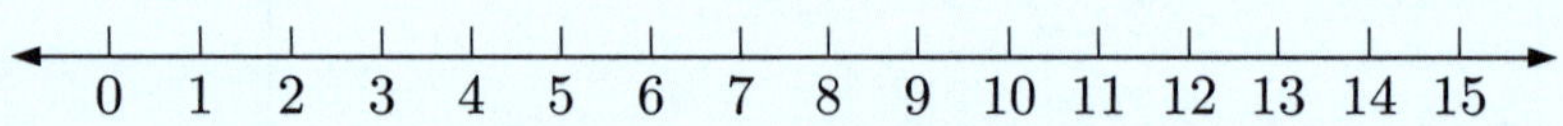

$13 - 6 =$ ______

d 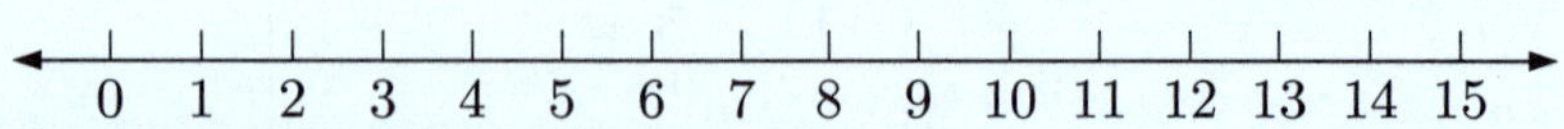

$15 - 3 =$ ______

e 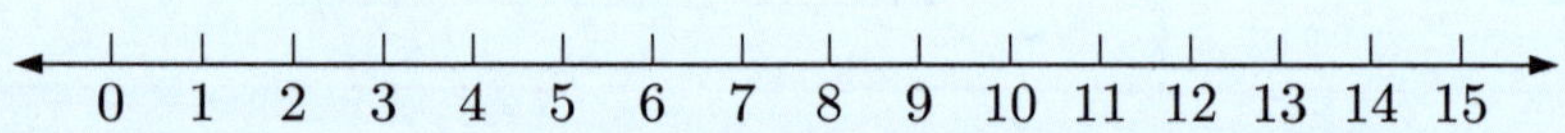

$14 - 9 =$ ______

Exercise 15

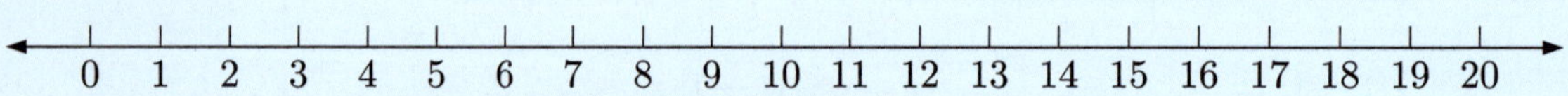

Use the number line to help with these subtractions.

a $15 - 4 =$ ______ **b** $17 - 3 =$ ______ **c** $14 - 2 =$ ______

d $13 - 3 =$ ______ **e** $12 - 5 =$ ______ **f** $13 - 7 =$ ______

g $16 - 6 =$ ______ **h** $18 - 8 =$ ______ **i** $17 - 9 =$ ______

Exercise 16

Use the addition to write *two* subtractions.

a $8 + 7 = 15$

so $15 -$ ______ $=$ ______

and $15 -$ ______ $=$ ______

b $7 + 9 =$ ______

so ______ $- 7 =$ ______

and ______ $- 9 =$ ______

Exercise 17

Complete each pattern of subtractions:

a

$10 - 10 = 0$

$20 - 10 = 10$

$30 - 10 =$ ......

$40 - 10 =$ ......

$50 - 10 =$ ......

$60 - 10 =$ ......

$70 - 10 =$ ......

$80 - 10 =$ ......

$90 - 10 =$ ......

b

$13 - 10 =$ ......

$23 - 10 =$ ......

$33 - 10 =$ ......

$43 - 10 =$ ......

$53 - 10 =$ ......

$63 - 10 =$ ......

$73 - 10 =$ ......

$83 - 10 =$ ......

$93 - 10 =$ ......

Activity

1 Start at 97 on the number chart. Colour it blue.

2 Count back 10 and colour the result green.

3 Repeat **2** to make a pattern.

1	2	3	4	5	6	7	8	9	10
11	12	13	14	15	16	17	18	19	20
21	22	23	24	25	26	27	28	29	30
31	32	33	34	35	36	37	38	39	40
41	42	43	44	45	46	47	48	49	50
51	52	53	54	55	56	57	58	59	60
61	62	63	64	65	66	67	68	69	70
71	72	73	74	75	76	77	78	79	80
81	82	83	84	85	86	87	88	89	90
91	92	93	94	95	96	97	98	99	100

Exercise 18

Complete these subtractions.

a $18 - 10 =$ ______ **b** $12 - 10 =$ ______ **c** $16 - 10 =$ ______

d $29 - 10 =$ ______ **e** $21 - 10 =$ ______ **f** $34 - 10 =$ ______

g $36 - 10 =$ ______ **h** $55 - 10 =$ ______ **i** $72 - 10 =$ ______

Puzzle

2	3	5	9

Use the cards to complete:

a ☐ + ☐ = 7

b ☐ − ☐ = 2

c ☐ − ☐ = 6

d ☐ + ☐ = ☐

Exercise 19

a Take away 2 each time:

13 → 11 → ☐ → ☐ → ☐

b Take away 2 each time:

14 → ☐ → ☐ → ☐ → ☐

c Take away 3 each time:

12 → ☐ → ☐ → ☐ → ☐

d Take away 3 each time:

16 → ☐ → ☐ → ☐ → ☐

e Take away 10 each time:

48 → ☐ → ☐ → ☐ → ☐

Exercise 20

Describe each sequence:

a 17 → 14 → 11 → 8 → 5

I start with ______, and subtract ______ each time.

b 16 → 12 → 8 → 4 → 0

I start with ______, and subtract ______ each time.

c 59 → 49 → 39 → 29 → 19

____

d 19 → 17 → 15 → 13 → 11

____

Exercise 21

Complete these subtractions.

a		**b**		**c**	
	$7 - 6 =$ ______		$7 - 7 =$ ______		$9 - 9 =$ ______
	$8 - 6 =$ ______		$8 - 7 =$ ______		$10 - 9 =$ ______
	$9 - 6 =$ ______		$9 - 7 =$ ______		$11 - 9 =$ ______
	$10 - 6 =$ ______		$10 - 7 =$ ______		$12 - 9 =$ ______
	$11 - 6 =$ ______		$11 - 7 =$ ______		$13 - 9 =$ ______
	$12 - 6 =$ ______		$12 - 7 =$ ______		$14 - 9 =$ ______
	$13 - 6 =$ ______		$13 - 7 =$ ______		$15 - 9 =$ ______
	$14 - 6 =$ ______		$14 - 7 =$ ______		$16 - 9 =$ ______
	$15 - 6 =$ ______		$15 - 7 =$ ______		$17 - 9 =$ ______

Find the missing number: $8 - \square = 3$

If I start with 8, how many should I take away, to leave 3?

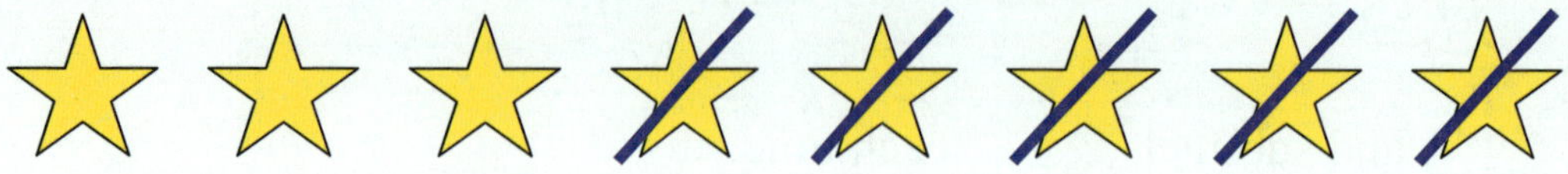

$8 - 5 = 3$

Exercise 22

Find the missing number.

a $6 - \square = 2$ **b** $5 - \square = 1$ **c** $8 - \square = 6$

d $10 - \square = 5$ **e** $9 - \square = 3$ **f** $13 - \square = 4$

g $15 - \square = 5$ **h** $17 - \square = 12$ **i** $16 - \square = 8$

Exercise 23

Using 2, 3, and 5, I can write:

$2 + 3 = 5$
$3 + 2 = 5$
$5 - 2 = 3$
$5 - 3 = 2$

a Use 1, 2, and 3 to complete:

$\square + \square = \square$
$\square + \square = \square$
$\square - \square = \square$
$\square - \square = \square$

b Use 3, 4, and 7 to complete:

$\square + \square = \square$
$\square + \square = \square$
$\square - \square = \square$
$\square - \square = \square$

Find the missing number: $\square - 5 = 3$

What do I need to start with, so when I take 5 away, I still have 3?

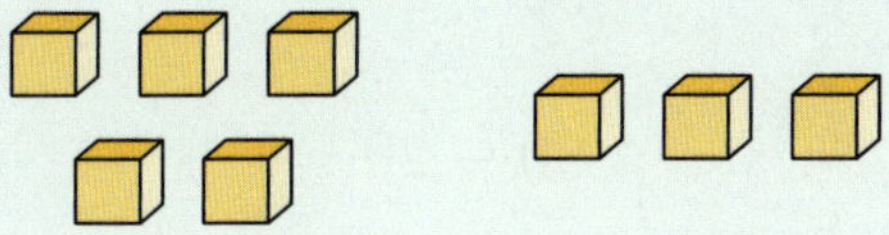

5 taken away 3 left

$8 - 5 = 3$

Exercise 24

Find the missing number.

a $\square - 2 = 2$ **b** $\square - 2 = 3$ **c** $\square - 1 = 4$

d $\square - 6 = 1$ **e** $\square - 4 = 3$ **f** $\square - 3 = 6$

g $\square - 5 = 7$ **h** $\square - 6 = 8$ **i** $\square - 3 = 9$

Exercise 25

Use the number line to help with these subtractions.

$20 - 0 =$ ______ $20 - 6 =$ ______

$20 - 1 =$ ______ $20 - 7 =$ ______

$20 - 2 =$ ______ $20 - 8 =$ ______

$20 - 3 =$ ______ $20 - 9 =$ ______

$20 - 4 =$ ______ $20 - 10 =$ ______

$20 - 5 =$ ______

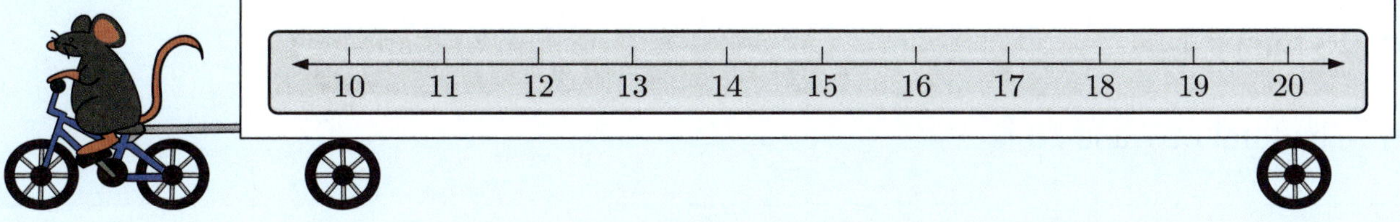

Exercise 26

Complete these subtractions.

a $60 - 10 =$ ______ **b** $70 - 10 =$ ______ **c** $90 - 10 =$ ______

$60 - 20 =$ ______ $70 - 20 =$ ______ $90 - 20 =$ ______

$60 - 30 =$ ______ $70 - 30 =$ ______ $90 - 40 =$ ______

$60 - 40 =$ ______ $70 - 40 =$ ______ $90 - 60 =$ ______

$60 - 50 =$ ______ $70 - 50 =$ ______ $90 - 80 =$ ______

Exercise 27

Write each word on the correct whiteboard:

sum total plus subtract

take away add minus

addition

subtraction

Exercise 28

Complete:

a 6 add 9 = ______

b 11 take away 6 = ______

c 14 subtract 7 = ______

d 11 plus 8 = ______

e 13 minus 4 = ______

f The sum of 4 and 8 is ______ .

g The total of 7 and 10 is ______ .

CHAPTER 6: MEASUREMENT

Discussion

We *measure* many things, including:

- length
- area
- volume
- capacity
- mass
- time

Discuss what each of these things mean.

Exercise 1

Practise your writing.

length

length

area

area

volume

volume

capacity

capacity

mass

mass

time

time

measurement

measurement

Exercise 2

If I want to compare the *size* of these objects, should I use *capacity*, *length*, or *area*?

a

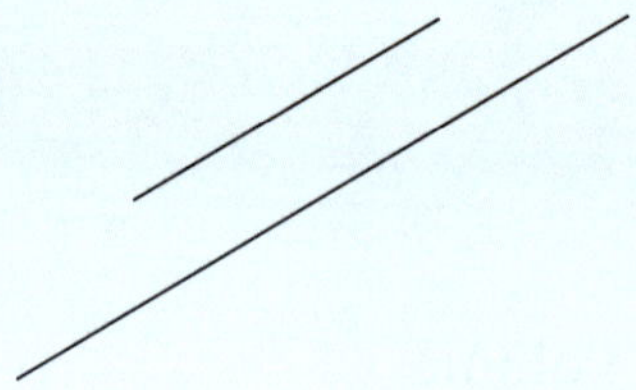

I can measure the ________________ of each line.

b

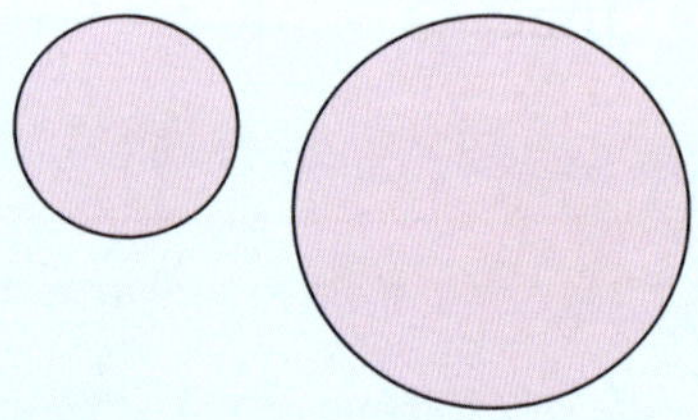

I can measure the ________________ of each circle.

c

I can measure the ________________ of each vase.

Discussion

Height, *width*, and *depth* are special words we use for ________________ measurements.

Why do you think we use them?

Hint: Think about measuring this fish tank.

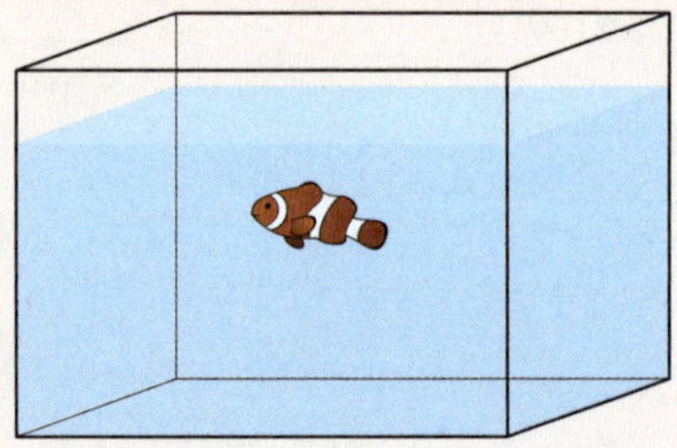

Exercise 3

Practise your writing.

height	width	depth
height	width	depth
____________	____________	____________

Exercise 4

What could we measure to *compare* these animals?

....

....

....

....

Activity

Discuss the meaning of each word. Copy it into the correct group.

little	narrow	big	large	long
short	tiny	massive	tall	light
huge	heavy	small	wide	

Length words	*Mass words*	*Area and volume words*

LENGTH

Exercise 5

earthwormmillipedecentipede

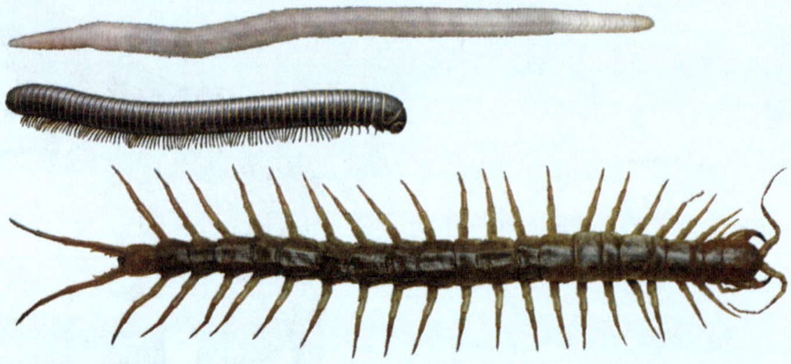

Complete using *longer*, *shorter*, *longest*, or *shortest*.

a The millipede is ....

b The centipede is ....

c The earthworm is .... than the millipede.

d The earthworm is .... than the centipede.

Exercise 6

Ross Allen Brandt Osgood

Write the squirrels in order of height.

....

shortest ⟶ tallest

Activity

Circle the words which describe a hippopotamus.

short wide tall

long narrow

Colour in the picture.

Activity

You will need: pencil, eraser, scissors, glue stick, exercise book, centicubes

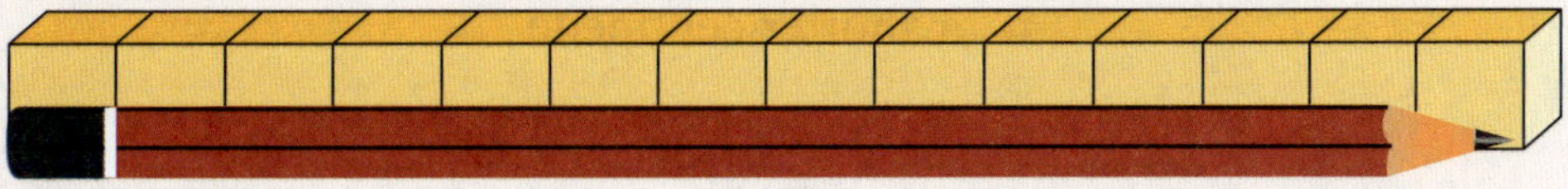

What to do:

1 Line up centicubes alongside each of your objects.

Record the length of each object in the table.

Object	*Length*
pencil	______ centicubes
eraser	______ centicubes
scissors	______ centicubes
glue stick	______ centicubes
exercise book	______ centicubes

2 The longest object was the ______________________.

The shortest object was the ______________________.

Exercise 7

Practise your writing.

shallow
shallow
____
____

deep
deep
____
____

Activity

Draw:

a a longer earthworm

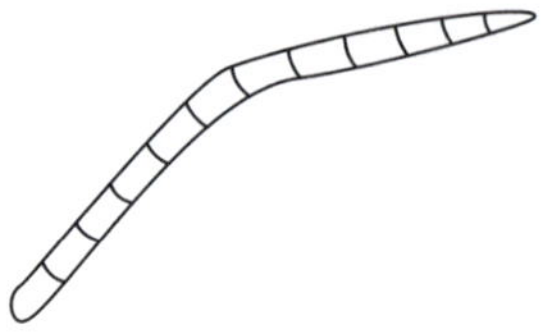

b a shorter centipede

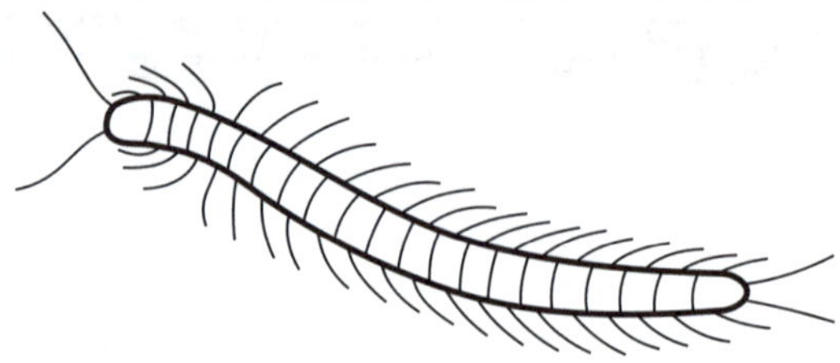

c a shorter person

d a taller building

e a pencil which is *shortest*

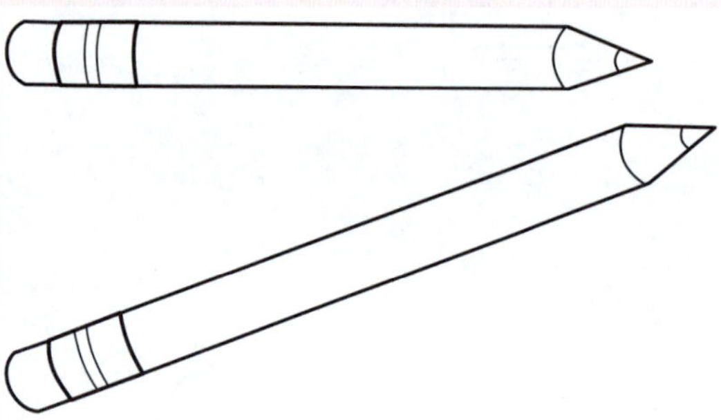

f a tree which is *tallest*.

Activity

How many cubes tall do you think you are? ____________________

Work in a small group to measure each person.

Complete: I am about ________ cubes tall.

Puzzle

Lucy is taller than Mary.

Hannah is taller than Mark but shorter than Mary.

Daniel is the shortest.

Draw the missing people.

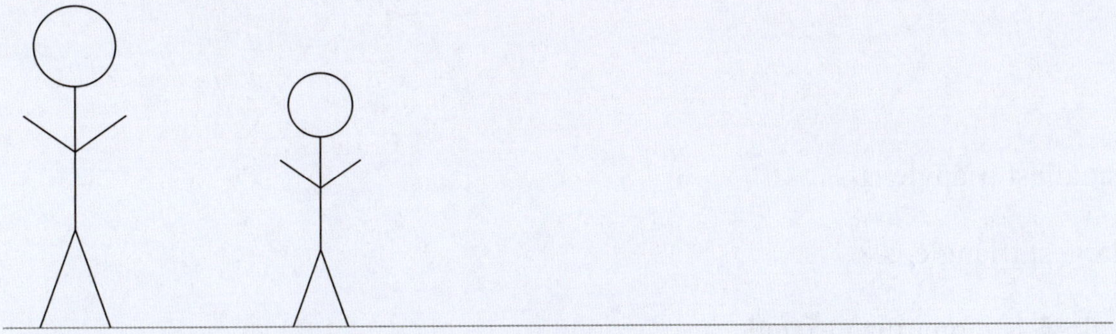

Mary Mark Lucy Hannah Daniel

AREA

The **area** of an object is the amount of *surface* it covers.

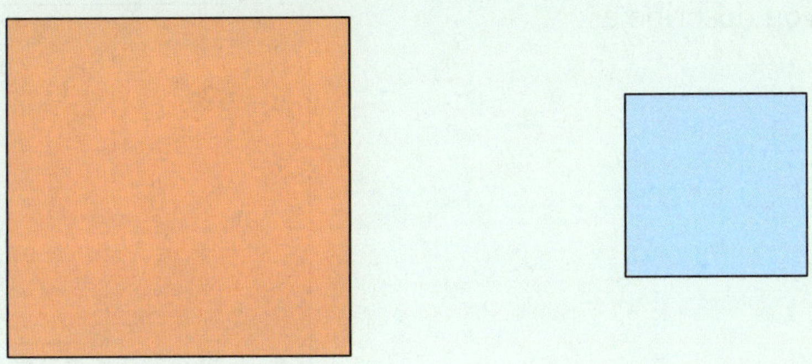

This square is *larger*.
It covers more area.

This square is *smaller*.
It covers less area.

Exercise 8

Use *small* and *large* to label each object.

a

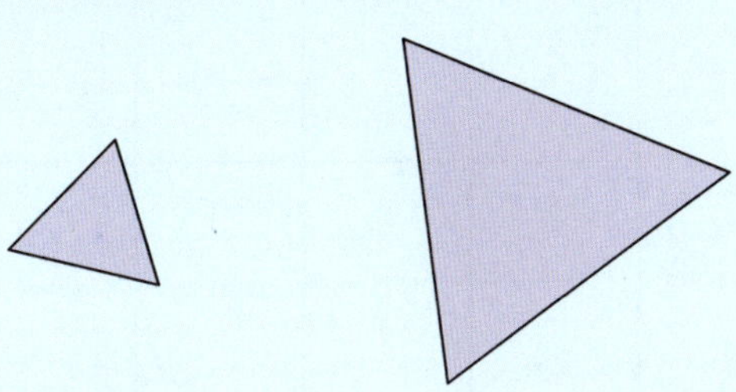

________ ________

b

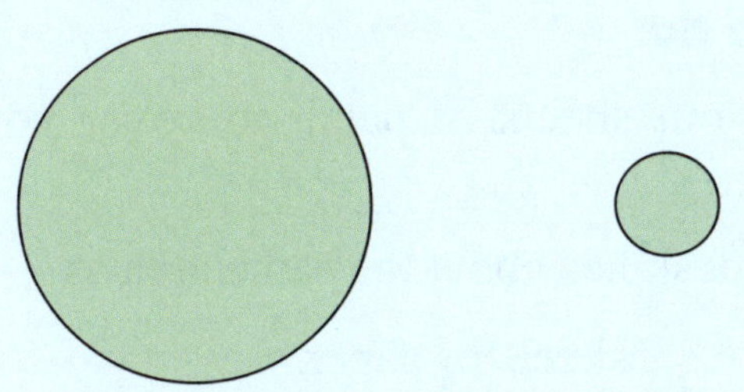

________ ________

Exercise 9

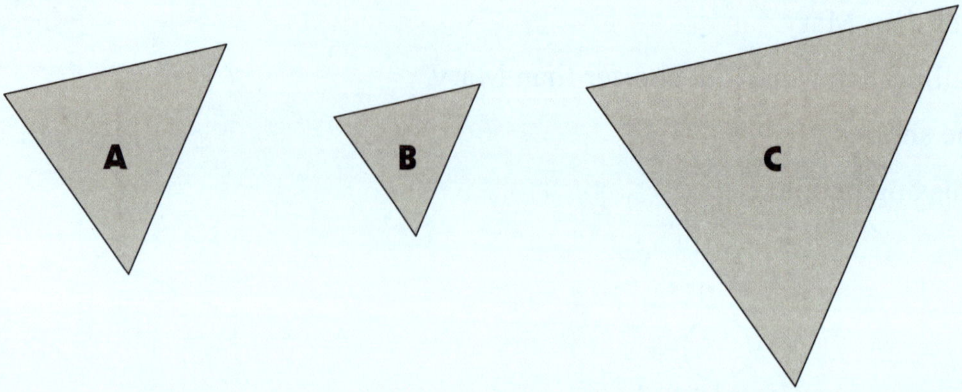

The smallest triangle is ______ .

The largest triangle is ______ .

Triangle **A** is larger than triangle ______ .

Triangle **B** is smaller than triangles ______ and ______ .

Exercise 10

Which shape would you describe as:

a little ____________________

b big ____________________

c tiny? ____________________

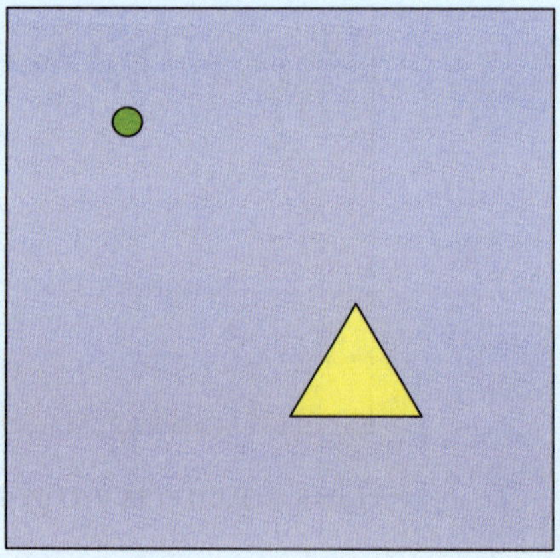

Activity

You will need: sheets of paper, scissors

What to do:

1 Lay out sheets of paper to cover your desk.

2 My desk has about the same area as ______ sheets of paper.

Activity

Draw:

a a larger rectangle

b a smaller circle

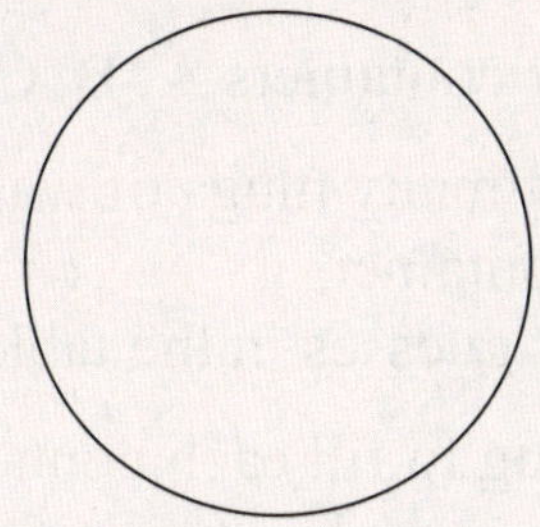

c a smaller triangle

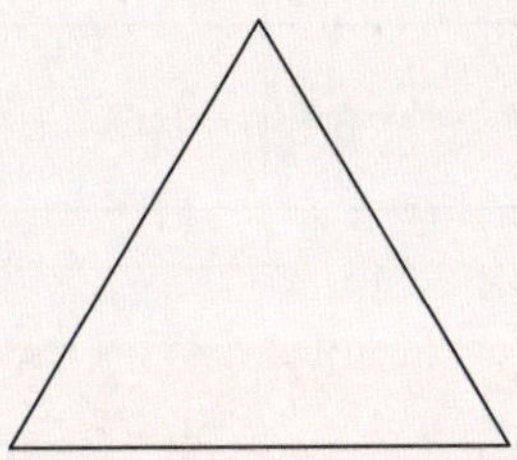

d a larger square.

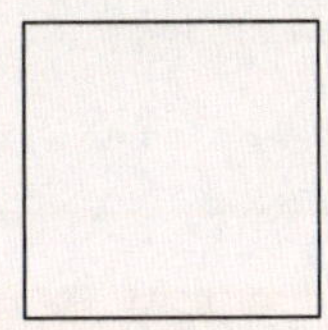

CAPACITY

The **capacity** of a container is the amount of liquid it can hold.

Exercise 11

Look at these containers.

glass

bucket

eye drop bottle

The container with the smallest capacity is the ______________________.

The container with the largest capacity is the ______________________.

Activity

You will need: 4 different containers, a mug

What to do:

1 Label your containers A, B, C, and D.

2 *Guess* how many mugs of water it will take to fill each container.
Write your guesses in the table.

3 Use the mug to fill each container.
Record the *actual* number of mugs needed.

4 The *smallest* container is ______ .

The *largest* container is ______ .

Container	*Guess*	*Actual*
A		
B		
C		
D		

Activity

Look around your classroom.

Write down two containers which have greater capacity than a mug.

1 ____________ **2** ____________

Write down two containers which have less capacity than a mug.

1 ____________ **2** ____________

Exercise 12

Look at the containers.

A **B** **C** **D**

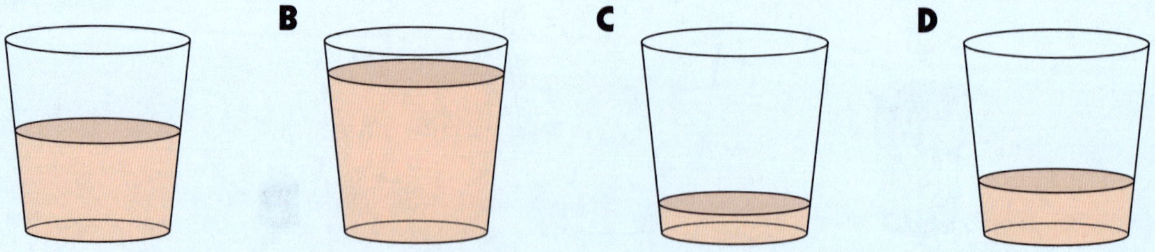

Write the amounts in order.

______ ______ ______ ______

least liquid → most liquid

Exercise 13

Practise your spelling.

empty

empty

full

full

Exercise 14

Match each bottle to the amount it contains.

full	almost full	empty	almost empty	half full

Exercise 15

Use *less* or *more* to compare the containers.

a

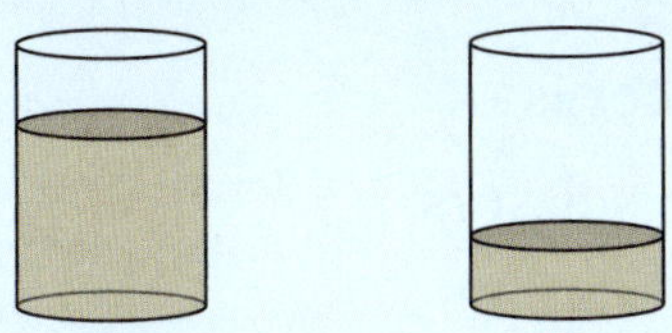

A **B**

A has ____________ than **B**.

b

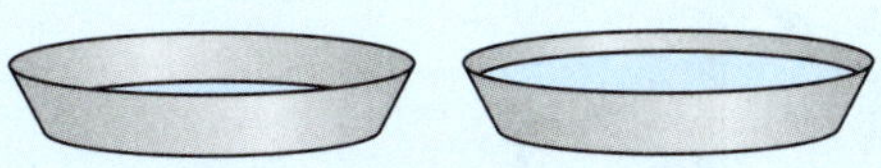

A **B**

A has ____________ than **B**.

c

A **B**

A has ____________ than **B**.

MASS

To compare mass, we can use a **balance scale**.

The pear is heavier than the grape.

Exercise 16

Use *heavier* or *lighter* to complete each sentence:

a

The pear is ....
than the cherry.

b

The eraser is ....
than the scissors.

c

The grape is ....
than the melon.

d

The squirrel is ....
than the acorn.

We *weigh* an object to find its mass.

Exercise 17

Complete using *less than*, *more than*, or *the same as*.

a

The apple weighs ______________ 4 blocks.

b

The orange weighs ______________ 4 blocks.

c

The egg weighs ______________ 4 blocks.

d

The carrot weighs ______________ 3 blocks.

e

The key weighs ______________ 3 blocks.

f

The book weighs ______________ 6 blocks.

Activity

You will need: balance scale, centicubes, glue stick, empty water bottle, tape, orange, tennis ball

What to do:

How many cubes will balance each item?

The spoon weighs 8 cubes.

Item	*Mass*
 glue stick	...... cubes
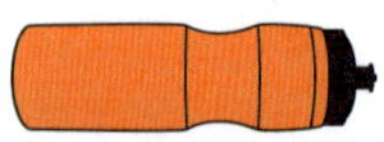 bottle	...... cubes
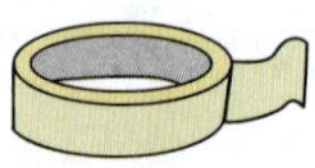 tape	...... cubes
 orange	...... cubes
 tennis ball	...... cubes

The heaviest item is the ....

The lightest item is the ....

CHAPTER 7: MULTIPLICATION

Here are 2 groups of 3 children.

In total, there are

$3 + 3 = 6$ children.

We can also write a **multiplication**:

$2 \times 3 = 6$

"two times three equals six"

"two multiplied by three is six"

Exercise 1

a Here are 3 groups of 2 children.

In total, there are

______ + ______ + ______ = ______ children.

As a multiplication, we write

______ $\times$ 2 = ______

b Here are 2 groups of 4 children.

In total, there are

______ + ______ = ______ children.

As a multiplication, we write

______ $\times$ 4 = ______

c Here are 4 groups of 2 children.

In total, there are

______ + ______ + ______ + ______ = ______ children.

As a multiplication, we write

______ $\times$ 2 = ______

Exercise 2

In this Exercise you will write your **2 times table**.

a

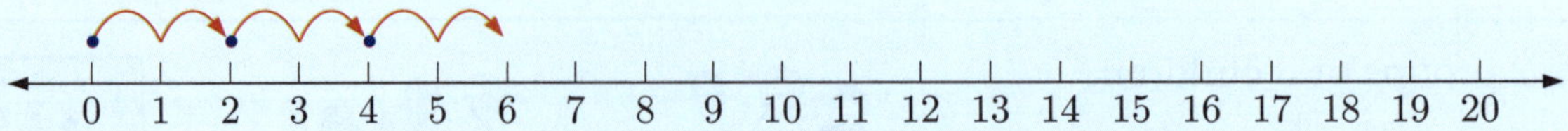

Start at 0 and add 2 each time.

0 → 2 → ______ → ______ → ______ → ______ → ______ → ______

→ ______ → ______ → ______

b A *pair* of glasses has *two* lenses.

Complete each multiplication:

1 pair is	1	$\times$	2	=	______
2 pairs is	2	$\times$	2	=	______
3 pairs is	______	$\times$	2	=	______
4 pairs is	______	$\times$	______	=	______
5 pairs is	______	$\times$	______	=	______
6 pairs is	______	$\times$	______	=	______
7 pairs is	______	$\times$	______	=	______
8 pairs is	______	$\times$	______	=	______
9 pairs is	______	$\times$	______	=	______
10 pairs is	______	$\times$	______	=	______

c Say your 2 times table together with your class.

Click the icon to practise by yourself.

Exercise 3

In this Exercise you will write your **5 times table**.

a

1	2	3	4	5	6	7	8	9	10
11	12	13	14	15	16	17	18	19	20
21	22	23	24	25	26	27	28	29	30
31	32	33	34	35	36	37	38	39	40
41	42	43	44	45	46	47	48	49	50

Start at 0 and add 5 each time.

0 → 5 → ____ → ____ → ____ → ____ → ____ → ____

→ ____ → ____ → ____

b Each foot has 5 toes.

Complete each multiplication:

1 foot has 1 × 5 = ____ toes.

2 feet have 2 × 5 = ____ toes.

3 feet have ____ × 5 = ____ toes.

4 feet have ____ × ____ = ____ toes.

5 feet have ____ × ____ = ____ toes.

6 feet have ____ × ____ = ____ toes.

7 feet have ____ × ____ = ____ toes.

8 feet have ____ × ____ = ____ toes.

9 feet have ____ × ____ = ____ toes.

10 feet have ____ × ____ = ____ toes.

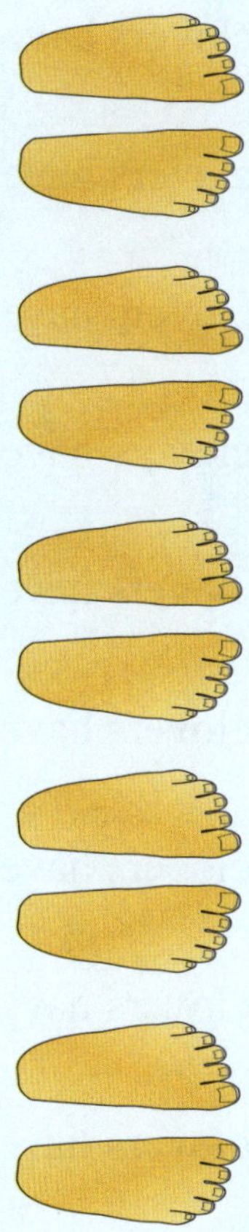

c Say your 5 times table together with your class.

Click on the icon to practise by yourself.

Exercise 4

In this Exercise you will write your **3 times table**.

a

1	2	3	4	5	6	7	8	9	10
11	12	13	14	15	16	17	18	19	20
21	22	23	24	25	26	27	28	29	30

Start at 0 and add 3 each time.

0 → 3 → ______ → ______ → ______ → ______ → ______ → ______

→ ______ → ______ → ______

b Each clover has 3 leaves.

Complete each multiplication:

1 clover has 1 × 3 = ______ leaves. 

2 clovers have 2 × 3 = ______ leaves. 

3 clovers have ______ × 3 = ______ leaves. 

4 clovers have ______ × ______ = ______ leaves. 

5 clovers have ______ × ______ = ______ leaves. 

6 clovers have ______ × ______ = ______ leaves. 

7 clovers have ______ × ______ = ______ leaves. 

8 clovers have ______ × ______ = ______ leaves. 

9 clovers have ______ × ______ = ______ leaves. 

10 clovers have ______ × ______ = ______ leaves. 

c Say your 3 times table together with your class.

Click on the icon to practise by yourself.

Exercise 5

In this Exercise you will write your **10 times table**.

a Start at 0 and add 10 each time.

0 → 10 → ______ → ______ → ______ → ______ → ______ → ______

→ ______ → ______ → ______

b Each person has 10 fingers.

Complete each multiplication:

1 person has	1	×	10	=	______	fingers.
2 people have	2	×	10	=	______	fingers.
3 people have	______	×	10	=	______	fingers.
4 people have	______	×	______	=	______	fingers.
5 people have	______	×	______	=	______	fingers.
6 people have	______	×	______	=	______	fingers.
7 people have	______	×	______	=	______	fingers.
8 people have	______	×	______	=	______	fingers.
9 people have	______	×	______	=	______	fingers.
10 people have	______	×	______	=	______	fingers.

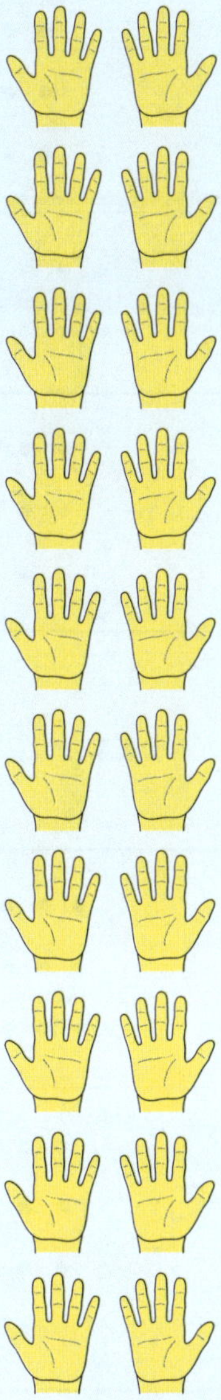

c Say your 10 times table together with your class.

Click on the icon to practise by yourself.

Exercise 6

Complete this table:

Picture	*Addition*	*Multiplication*	*Answer*
	$2+2+2+2+2$	5×2	10
	$3+3+3$		
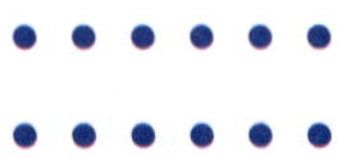	$6+6$		
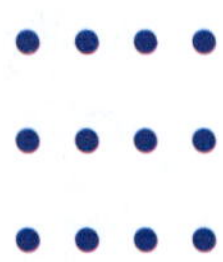		3×4	
		2×5	
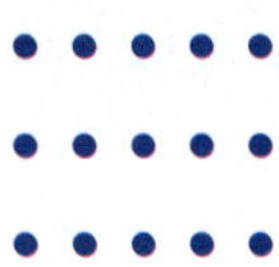			
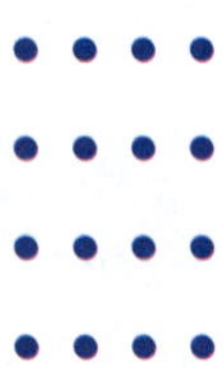			
	$3+3+3+3$		

Exercise 7

Complete this 1 times table.

$1 \times 1 =$ ______ $6 \times 1 =$ ______

$2 \times 1 =$ ______ $7 \times 1 =$ ______

$3 \times 1 =$ ______ $8 \times 1 =$ ______

$4 \times 1 =$ ______ $9 \times 1 =$ ______

$5 \times 1 =$ ______ $10 \times 1 =$ ______

Discussion

If we multiply any number by 1, what happens to the number?

____

Exercise 8

Complete these times tables.

a

$1 \times 2 =$ ______

$2 \times 2 =$ ______

$3 \times 2 =$ ______

$4 \times 2 =$ ______

$5 \times 2 =$ ______

$6 \times 2 =$ ______

$7 \times 2 =$ ______

$8 \times 2 =$ ______

$9 \times 2 =$ ______

$10 \times 2 =$ ______

b

$1 \times 3 =$ ______

$2 \times 3 =$ ______

$3 \times 3 =$ ______

$4 \times 3 =$ ______

$5 \times 3 =$ ______

$6 \times 3 =$ ______

$7 \times 3 =$ ______

$8 \times 3 =$ ______

$9 \times 3 =$ ______

$10 \times 3 =$ ______

c

$1 \times 5 =$ ______

$2 \times 5 =$ ______

$3 \times 5 =$ ______

$4 \times 5 =$ ______

$5 \times 5 =$ ______

$6 \times 5 =$ ______

$7 \times 5 =$ ______

$8 \times 5 =$ ______

$9 \times 5 =$ ______

$10 \times 5 =$ ______

Exercise 9

The number 40 means 4 tens or 4×10.

$4 \times 10 = 40$

Complete:

$1 \times 10 =$ ______	$6 \times 10 =$ ______
$2 \times 10 =$ ______	$7 \times 10 =$ ______
$3 \times 10 =$ ______	$8 \times 10 =$ ______
$4 \times 10 =$ ______	$9 \times 10 =$ ______
$5 \times 10 =$ ______	$10 \times 10 =$ ______

Discussion

What do you notice about the results in the 10 times table?

____

Exercise 10

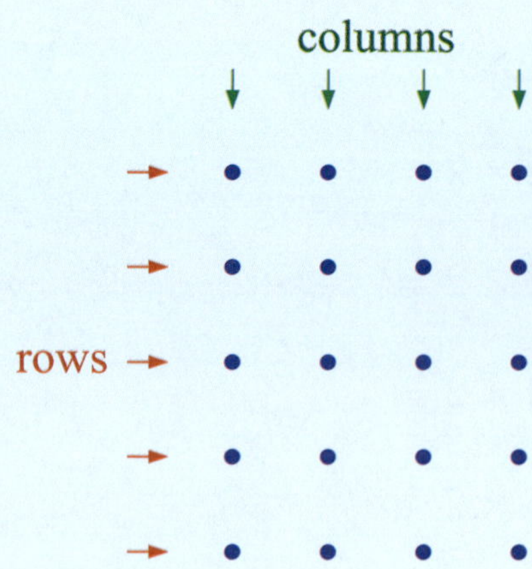

a There are 5 rows with 4 dots in each row.

Complete:

$4 + 4 + 4 + 4 + 4 =$ ______

$5 \times 4 =$ ______

b There are 4 columns with 5 dots in each column.

Complete:

$5 + 5 + 5 + 5 =$ ______

$4 \times 5 =$ ______

Discussion

What do you notice about the multiplications in **Exercise 10**?

____

CHAPTER 8: DATA HANDLING

When we look at the world around us, we often **count** things.
The information we collect is called **data**.

Exercise 1

Sharon counted these birds from her window.

Bird	*Number*
sparrow	3
raven	1
robin	4
pigeon	2

a How many sparrows did Sharon see? ______

b How many pigeons did Sharon see? ______

c Which bird did Sharon see the most of?

____

Exercise 2

Mr Harris has these balls in the sports shed.

a Mr Harris has:

☐ basketballs

☐ footballs

☐ baseballs.

b Mr Harris has ☐ more baseballs than footballs.

c Mr Harris buys 3 new basketballs.

He now has ☐ basketballs.

Bianca counts the colours of the pegs on her line.

She uses a **tally**.

She draws one **tally mark** for each item.

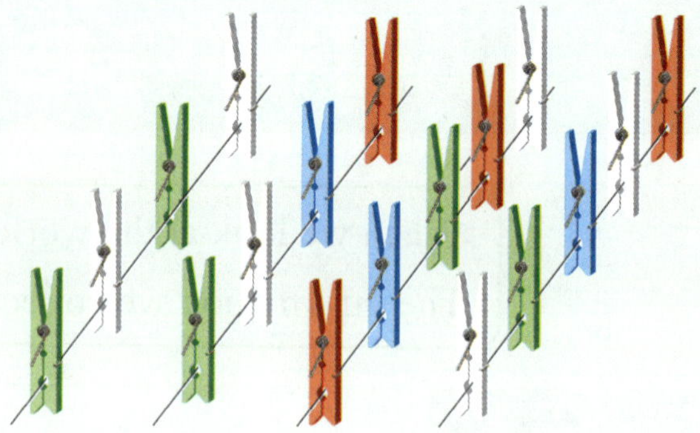

𝍷 represents 1 item.

𝍸 represents 5 items.

Colour	*Tally*
red	𝍷𝍷𝍷𝍷
white	𝍸 𝍷
blue	𝍷𝍷𝍷
green	𝍸

Bianca counted: 4 red pegs
6 white pegs
3 blue pegs
5 green pegs

Exercise 3

Mrs Jacobs asked her class "Where do you like to spend your lunch time?"

Place	*Tally*
Library	𝍷𝍷𝍷𝍷
Oval	𝍸 𝍷𝍷𝍷
Courts	𝍸
Courtyard	𝍷𝍷𝍷
Playground	𝍸 𝍷𝍷

a How many students chose the courtyard? ______

b How many students chose the playground? ______

c 4 students chose the ______________.

d The most popular place was the ______________.

Exercise 4

Count these animals using a tally.

Animal	*Tally*
bee	
frog	
grasshopper	

There are ☐ bees, ☐ frogs, and ☐ grasshoppers.

There are ☐ animals in total.

Exercise 5

Count these items using a tally.

Item	*Tally*
cup	
glass	
mug	

There are ☐ cups, ☐ glasses, and ☐ mugs.

There are ☐ items in total.

Group Activity

You will need:

- a printout of a puppy from your teacher
- coloured pencils including black, grey, brown, and yellow.

What to do:

1 Choose *one* colour you would like your puppy to be.
Colour your puppy in that colour.

2 Colour the rest of the picture as you wish.

3 Give your picture to your teacher.

4 Your teacher will show the pictures to the class one at a time.
Count the puppy colours using a tally:

Colour	*Tally*
black	
grey	
brown	
yellow	

5 The most popular colour chosen was ....

6 The least popular colour chosen was ....

Activity — What can you see?

a What animals can you see? ....

....

b Record how *many* of each animal you can see with a tally.

Animal	*Tally*

c Which animal can you see the most of? ....

d Which animal can you see the least of? ....

Exercise 6

Daniel counted the shirts in his wardrobe.

T-shirt polo shirt short-sleeved shirt long-sleeved shirt

He drew a **pictograph** to show his data.

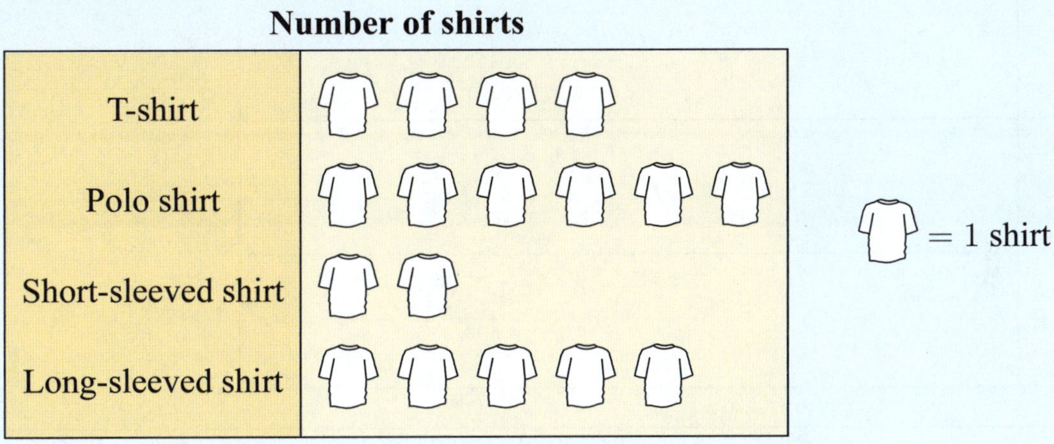

a How many long-sleeved shirts does Daniel have? ....

b How many T-shirts does Daniel have? ....

c Which type of shirt does Daniel have the most of? ....

d How many more polo shirts than short-sleeved shirts does Daniel have? ....

e In total, how many shirts does Daniel have in his wardrobe? ....

Discussion

What features of the pictograph above help us *understand* what it means?

....

....

....

Exercise 7

Delilah counted the trees in a garden.

Draw a pictograph to show this information.

Use  = 1 tree

Trees

Birch	6
Oak	4
Pine	5
Willow	2

Number of trees

Birch	
Oak	
Pine	
Willow	

Exercise 8

Nick counted the books in his bookcase.

Draw a pictograph to show this information.

Use  = 1 book

Book types

Animal	𝍷𝍷𝍷
Story	𝍸 𝍷
Joke	𝍷𝍷
Picture	𝍸

Number of books

Animal	
Story	
Joke	
Picture	

CHAPTER 9: DIVISION

Sita has 10 lychees. She wants to share them **equally** with Ruhi.

Sita's share

Ruhi's share

Each person will have 5 lychees.

Sita has **divided** the lychees **equally** into 2 groups.

Discussion

What do we mean by sharing *equally*?

Exercise 1

Describe the division:

a 

4 has been divided into ______ equal groups of ______ .

b 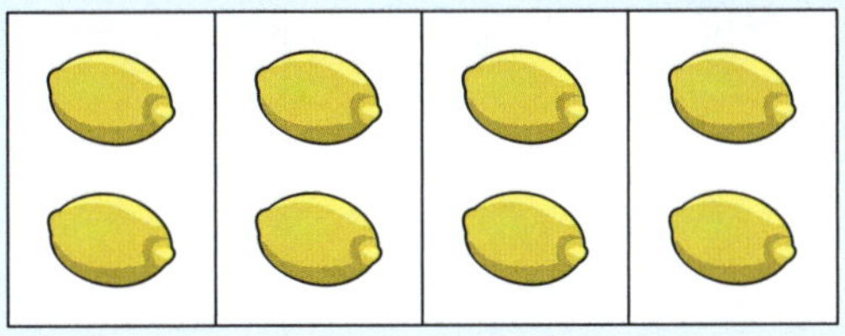

8 has been divided into ______ equal groups of ______ .

c 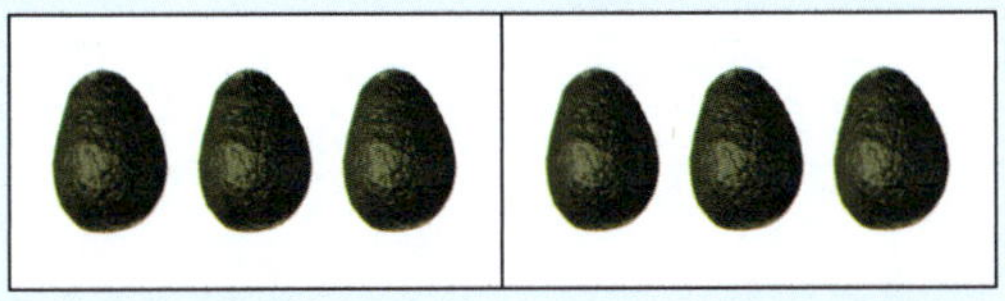

______ has been divided into ______ equal groups of ______ .

d 

______ has been divided into ______ equal groups of ______ .

Exercise 2

Describe the division:

a 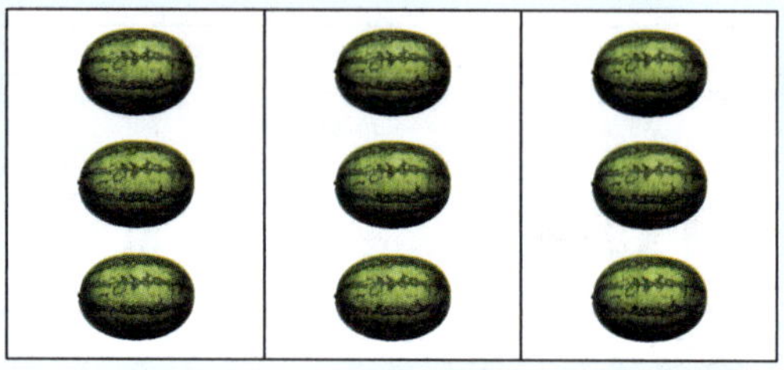

______ has been divided into ______ equal groups of ______ .

b 

______ has been divided into ______ equal groups of ______ .

Activity

You will need: 12 counters

What to do:

1 Divide the counters into 3 equal groups.

Draw your results.

12 has been divided into 3 equal groups of ______ .

2 Divide the counters into 2 equal groups.

Draw your results.

12 has been divided into 2 equal groups of ______ .

Exercise 3

Pip has 14 chocolates. She shares them equally with Mike.

Pip gives ______ chocolates to Mike.

Exercise 4

Practise your writing.

division

division

______

______

divide

divide

______

______

Angela cut her orange into 2 equal pieces.

She has **divided** her orange into 2.

Each piece is **one half** of the orange.

The whole orange has been divided into 2 halves.

Exercise 5

Practise your writing.

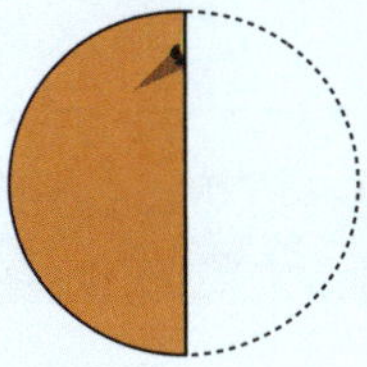

one half

one half

______

______

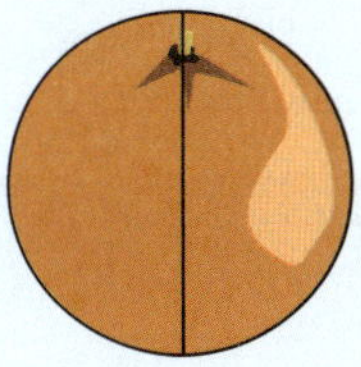

two halves

two halves

______

______

Exercise 6

Shade one half of each shape:

a

b

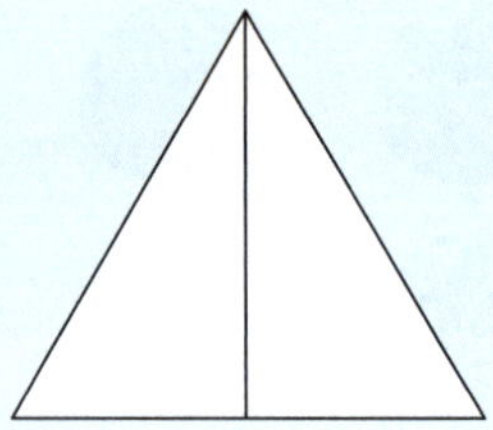

c

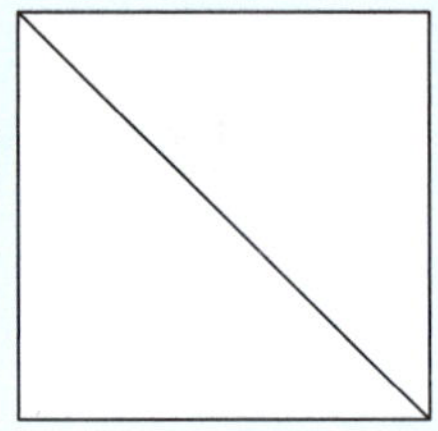

d

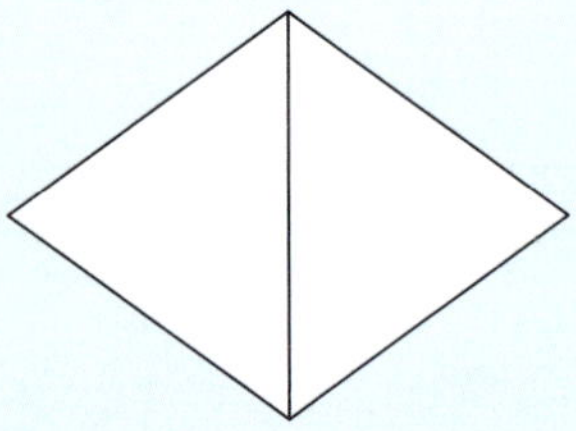

e

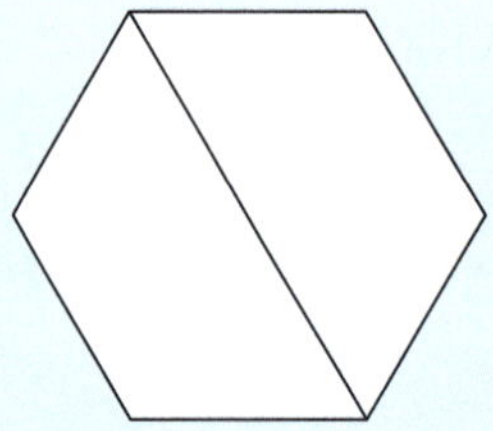

f

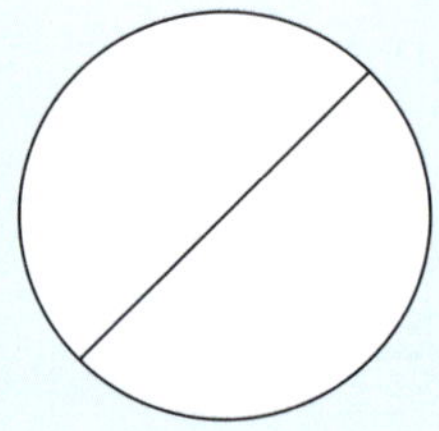

Exercise 7

Complete:

To divide a shape into *halves*, we divide it into ______ ____________ pieces.

There are ______ halves in one whole.

Exercise 8

How many halves are shaded?

a

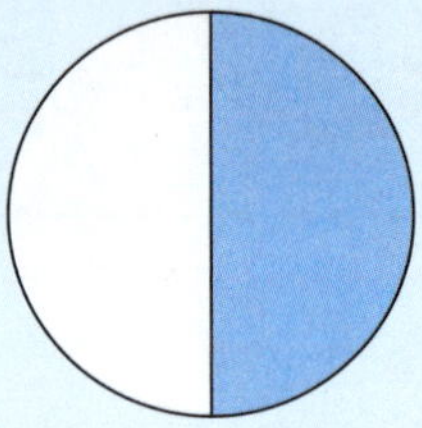

______ half

b

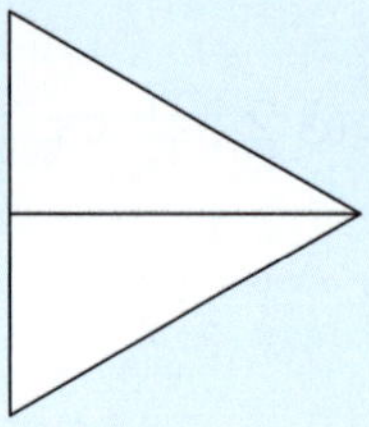

______ halves

c

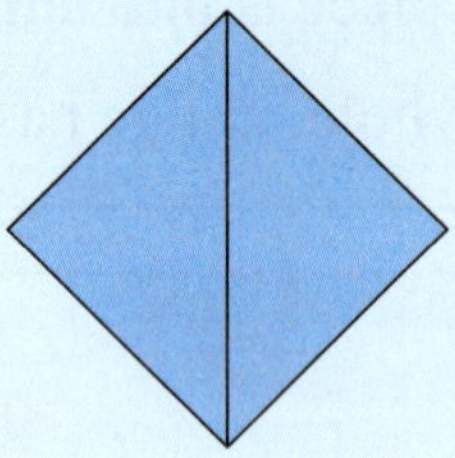

______ halves

Discussion

Why do these shapes *not* show one half?

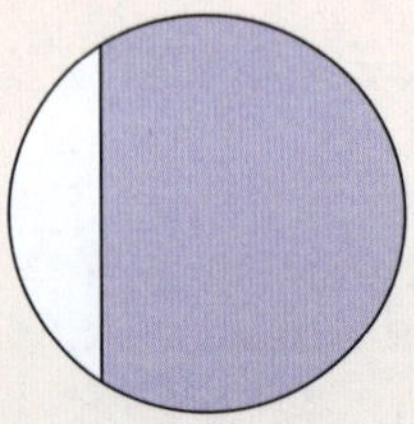

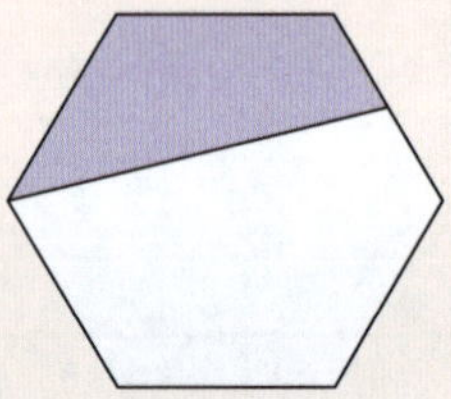

Exercise 9

Shade one half of each shape:

a 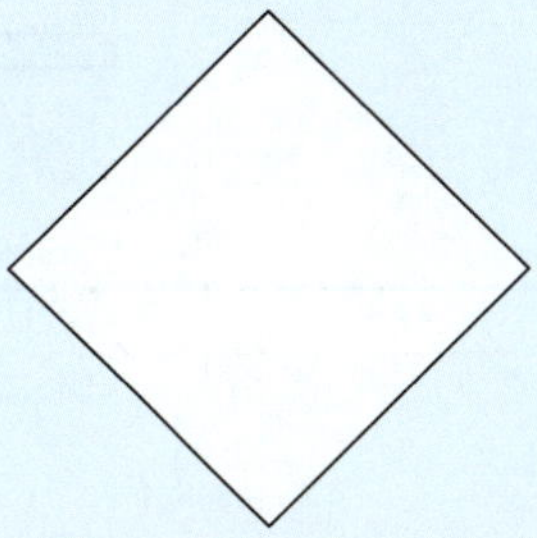

b 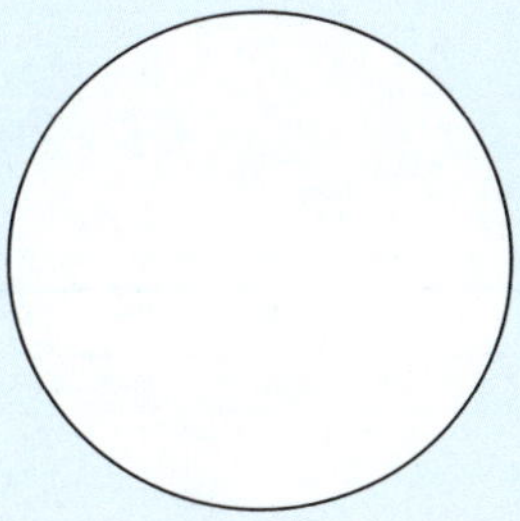

c 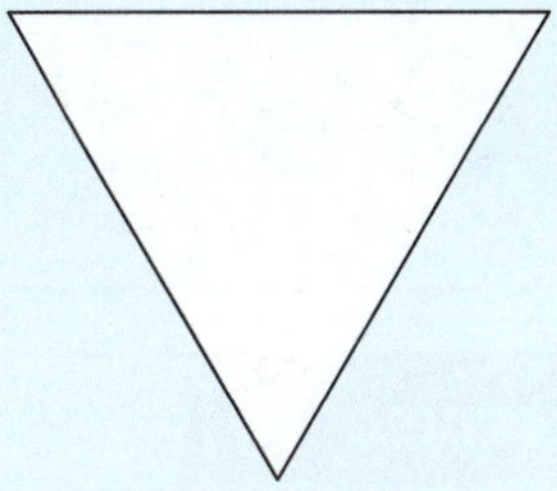

Activity

You wil need:

Printed shapes, scissors, coloured pencils, ruler.

What to do:

1 Cut out each shape.

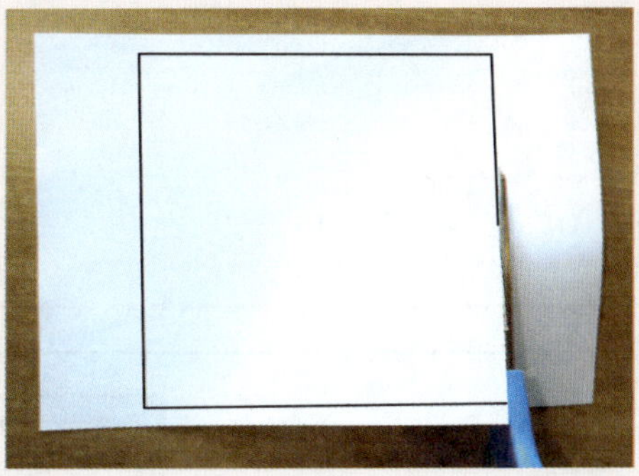

2 Carefully fold the square in half.
Draw a red line along the fold.

3 Fold the square in half a different way.
Draw a green line along the fold.

4 How else can you fold the square in halves?
Draw coloured lines along each fold.

5 Fold and mark halves on the other shapes.

To find **one half** of a number, we divide it into **two equal groups**.

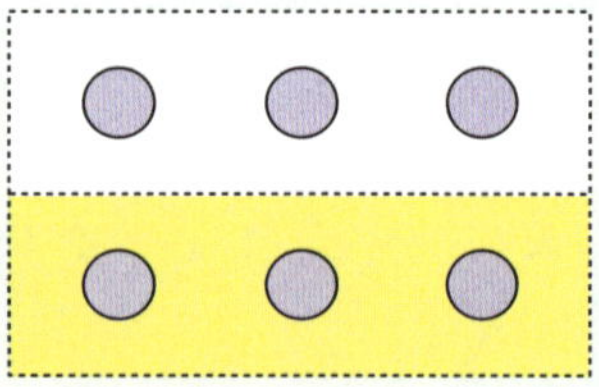

One half of 6 is 3.

Exercise 10

Circle one half of the objects.

a

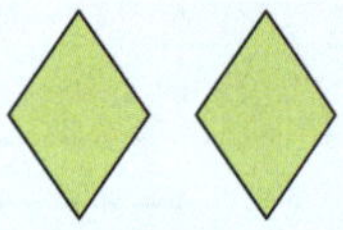

One half of ______ is ______ .

b

One half of ______ is ______ .

c

One half of ______ is ______ .

d

One half of ______ is ______ .

Jeff cut a log into 4 equal pieces.

He has **divided** the log into 4.

Each piece is **one quarter** of the log.

Exercise 11

Practise your writing.

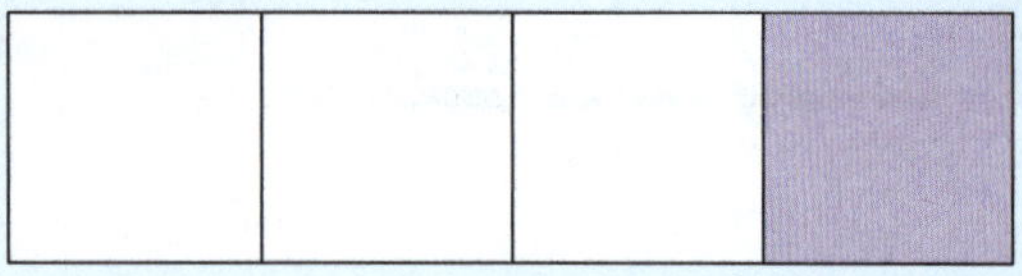

one quarter

one quarter

____

____

Exercise 12

Shade one quarter of each shape:

a
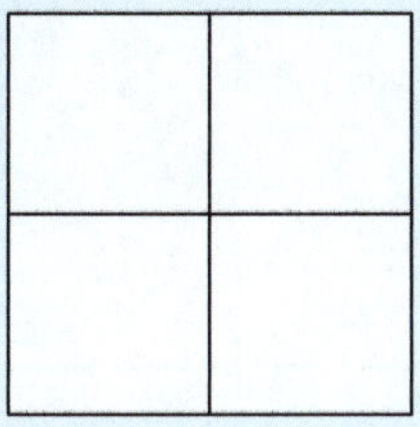

b
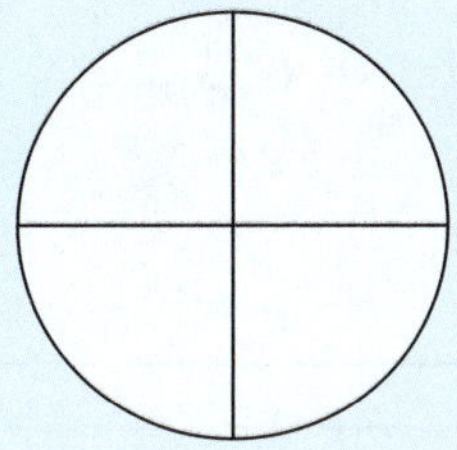

c
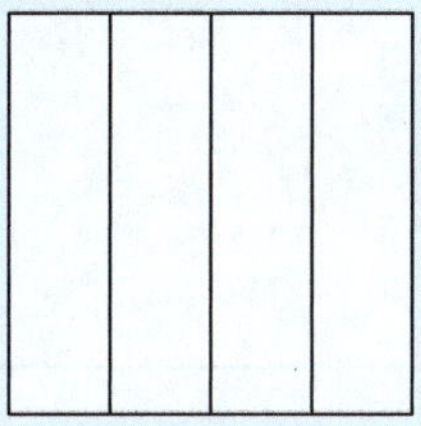

Exercise 13

Complete:

To divide a shape into *quarters*, we divide it into ______ __________ pieces.

There are ______ quarters in one whole.

Exercise 14

How many quarters are shaded?

a
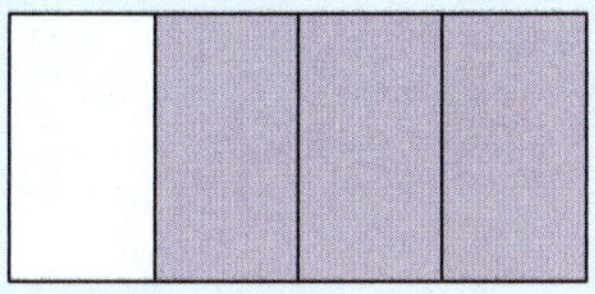

______ quarters

b
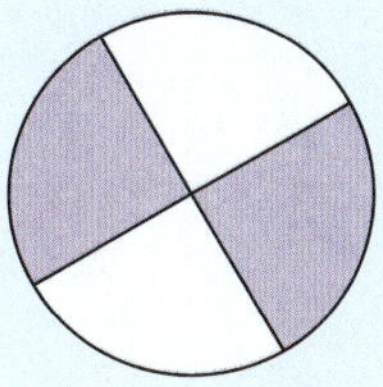

______ quarters

c
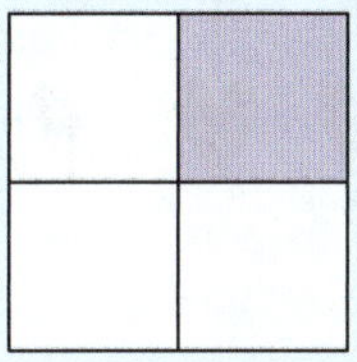

______ quarter

Exercise 15

Shade the correct amount.

a one quarter
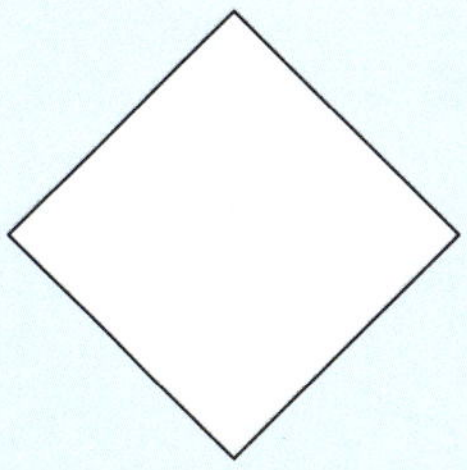

b one quarter
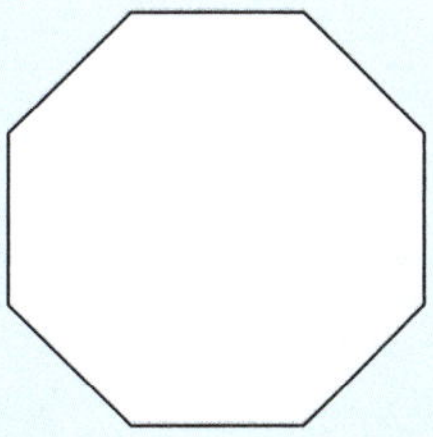

c three quarters
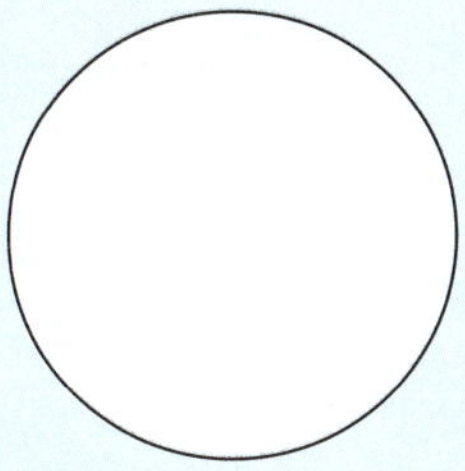

d one quarter
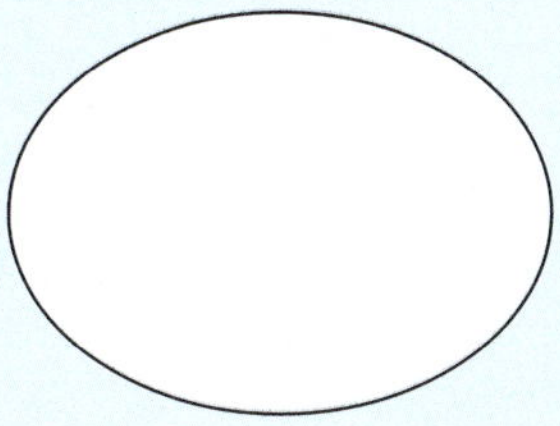

e two quarters

f three quarters
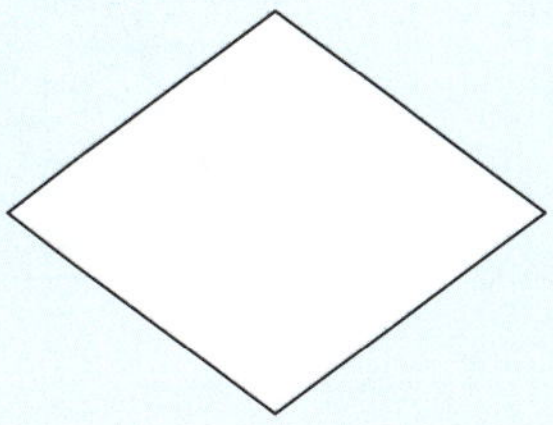

To find **one quarter** of a number, we divide it into **four equal groups**.

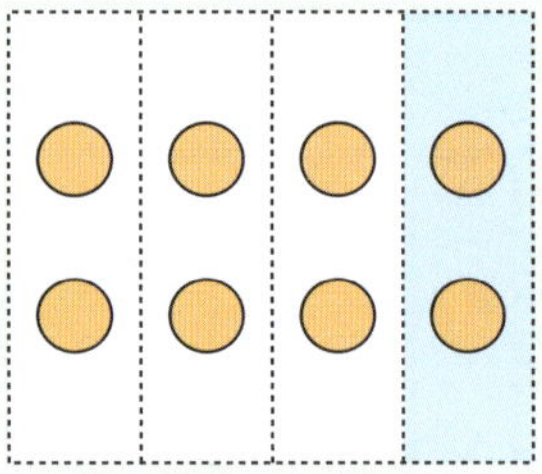

One quarter of 8 is 2.

Exercise 16

Circle one quarter of the objects.

a

One quarter of ______ is ______ .

b

One quarter of ______ is ______ .

c

One quarter of ______ is ______ .

d

One quarter of ______ is ______ .

Exercise 17

How much is shaded?

a

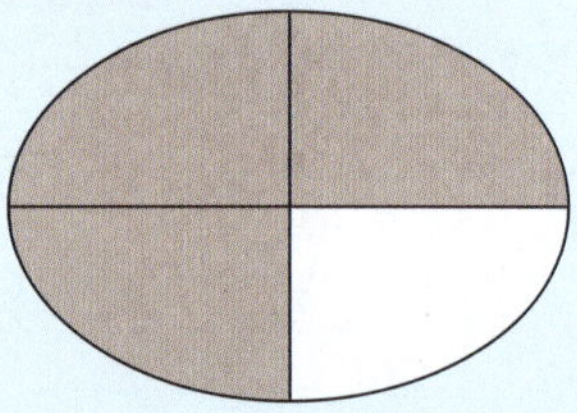

____

b

____

c

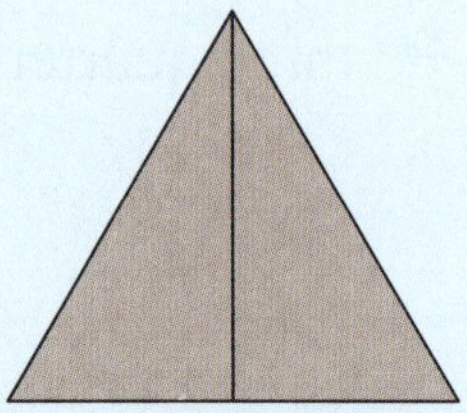

____

d

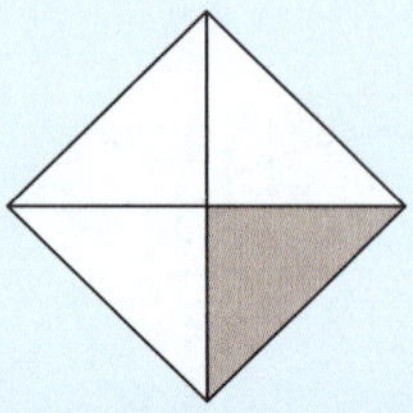

____

CHAPTER 10: TIME

Time measures *when* things happen.

We cook dinner **before** we eat it.

We go to school **after** we get dressed.

Exercise 1

Practise your writing.

a time

time

b before

before

c after

after

Today is the day right now.

Yesterday is the day *before* today.

Tomorrow is the day *after* today.

Exercise 2

Practise your writing.

a today

today

b tomorrow

tomorrow

c yesterday

yesterday

Exercise 3

Circle what you do *first* each day.

a
get dressed
wake up
brush teeth

b
eat lunch
go to bed
eat breakfast

c
go to sleep
brush teeth
wake up

Exercise 4

Circle what you do *last* each day.

a
have a bath
play with friends
bedtime story

b
wake up
put on pyjamas
eat breakfast

c
eat dinner
go to school
brush teeth

Discussion

What do we mean by:

- early
- on time
- late?

Exercise 5

Practise your writing.

early
early

late
late

We use a **clock** to tell the time.

The **minute hand** is longer. It is pointing to 12.

The **hour hand** is shorter. It is pointing to 4.

The time is 4 **o'clock**.

The hands on a clock always **turn** in the same direction.

The hands turn **clockwise**.

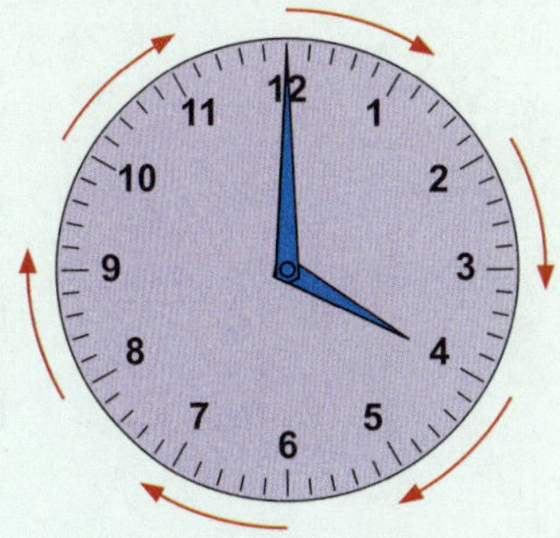

Exercise 6

Practise your writing.

a minute

minute

______

______

b hour

hour

______

______

c o'clock

o'clock

______

______

Exercise 7

What time is shown on the clock?

a

______ o'clock

b

______ o'clock

c

______ o'clock

d

______

e

______

f

______

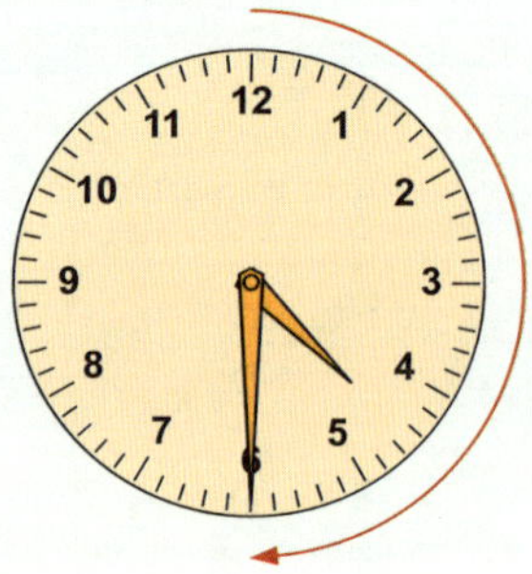

The minute hand has turned from 12 to 6.

The minute hand is **half way** around the clock.

The hour hand is **half way** between 4 and 5.

The time is **half past** 4.

Exercise 8

The minute hand is ______________ around the clock.

The hour hand is ______________ between ______ and ______ .

The time is ______________ .

Exercise 9

What time is shown on the clock?

a

______ past ______

b

______ past ______

c

______ past ______

d

____

e

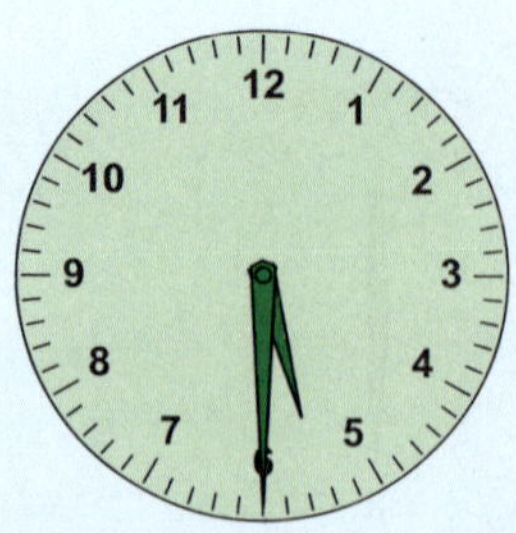

____

f

____

Exercise 10

a Label the hands of the clock.

b Draw the clockwise direction.

c The time is ....

....

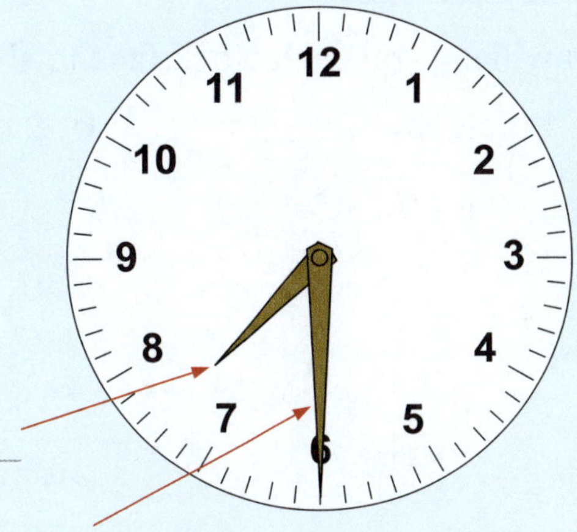

Exercise 11

Draw the missing hour hand.

a 1 o'clock

b 5 o'clock

c 11 o'clock

d half past 2

e half past 6

f half past 9

Exercise 12

Draw the missing minute hand.

a 9 o'clock

b half past 7

c half past 10

Exercise 13

Draw hands on each clock face to show the time:

a 4 o'clock

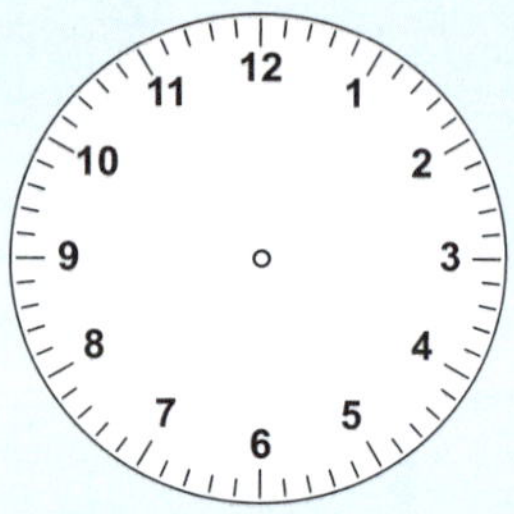

b 9 o'clock

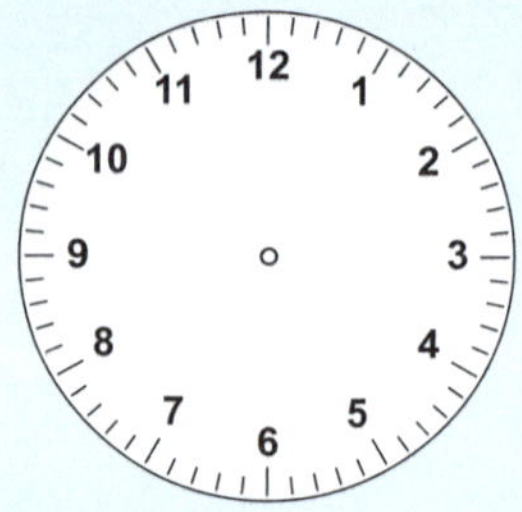

c 12 o'clock

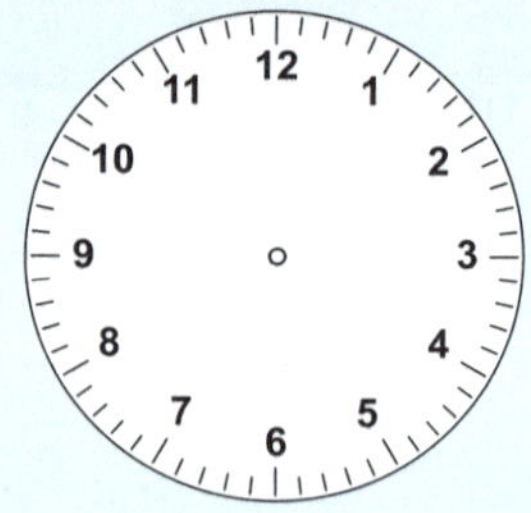

d half past 2

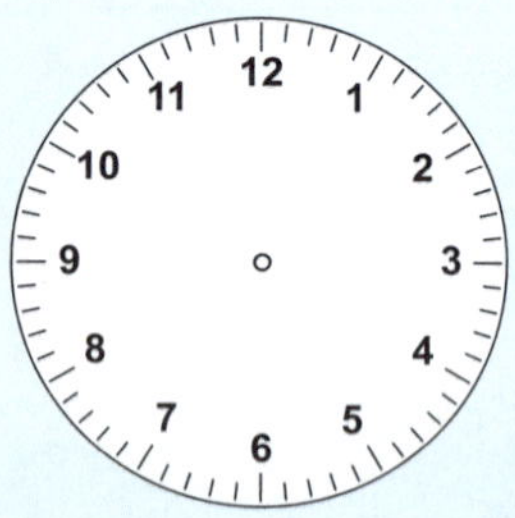

e half past 5

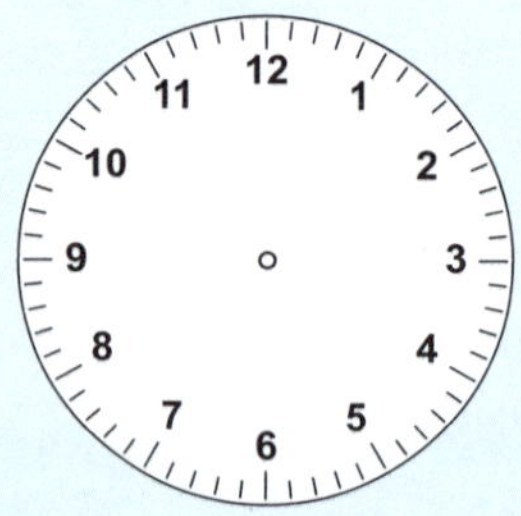

f half past 11

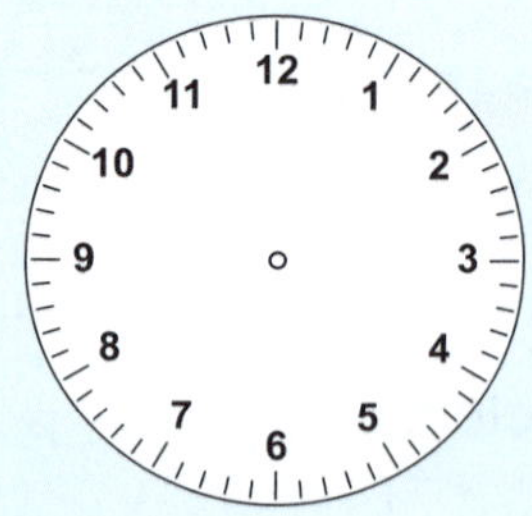

g 7 o'clock

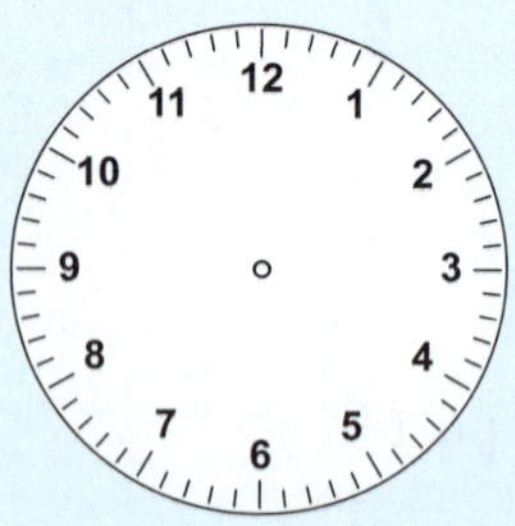

h half past 8

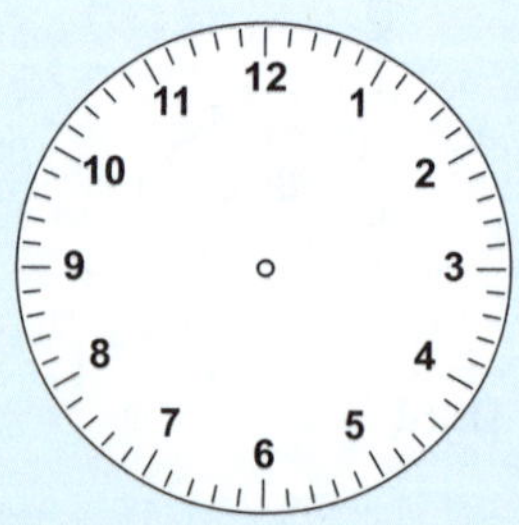

i half past 12

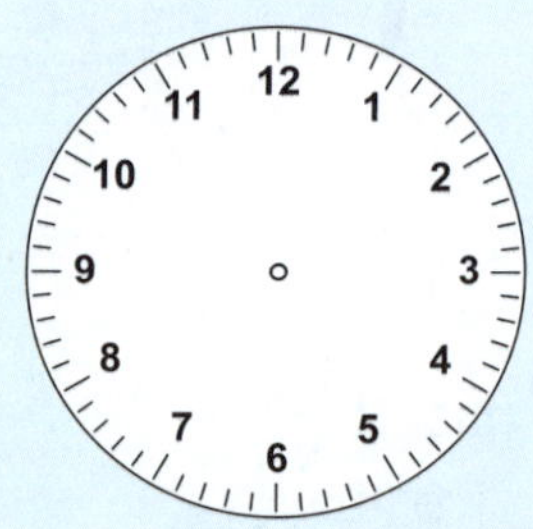

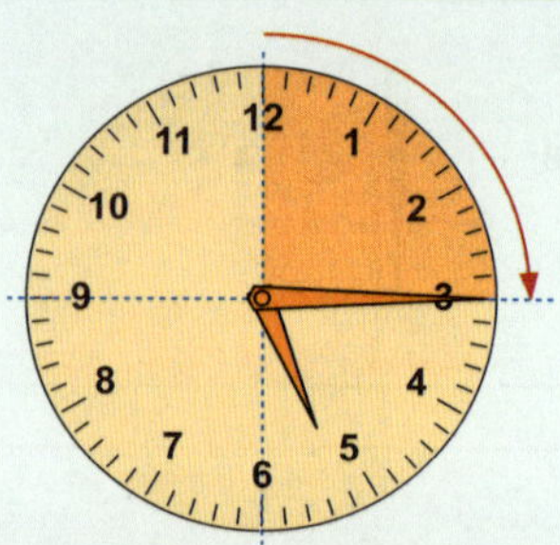

The minute hand has turned from 12 to 3.

The minute hand is a **quarter** of the way around the clock.

The hour hand is between 5 and 6.

The time is **quarter past** 5.

Exercise 14

The minute hand is a ________________ of the way around the clock.

The hour hand is between ______ and ______.

The time is ____________________.

Exercise 15

What time is shown?

a

quarter past ______

b

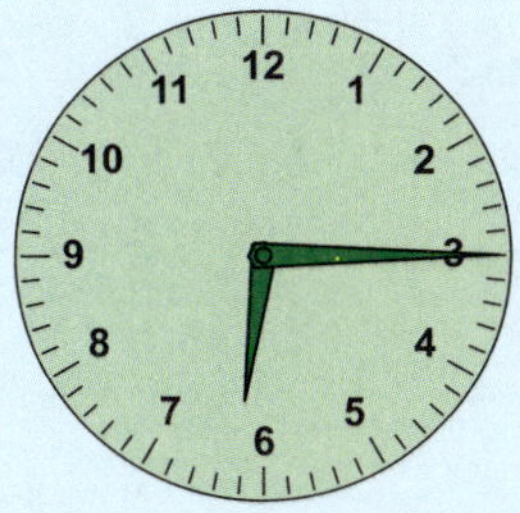

quarter past ______

c

quarter past ______

d

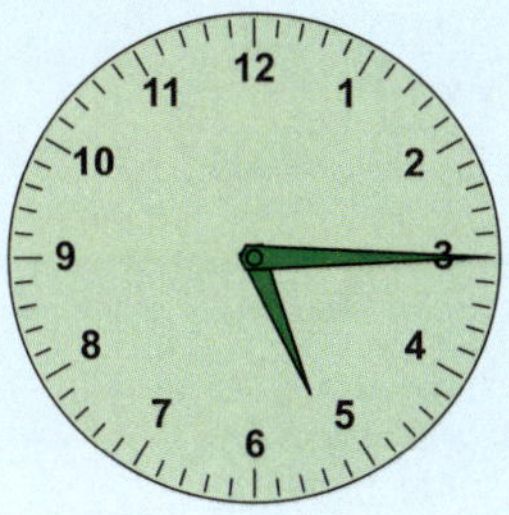

____

past ______

e

____

past ______

f

____

____

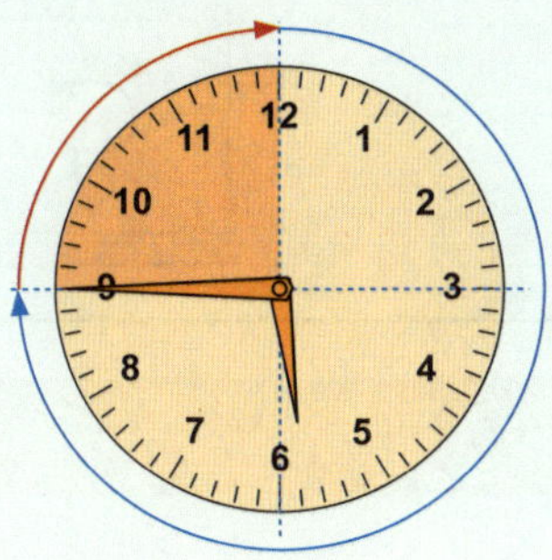

The minute hand has turned from 12 to 9.

The minute hand still has a **quarter** of the way to go, to get back to 12.

The hour hand is between 5 and 6.

The time is **quarter to** 6.

Exercise 16

The minute hand still has a ______________ of the way to go, to get back to 12.

The hour hand is between ______ and ______.

The time is ______________________.

Exercise 17

What time is shown?

a

quarter to ______

b

quarter to ______

c

quarter to ______

d

____

to ______

e

____

to ______

f

____

____

There are 60 minutes in **one hour**.

There are 24 hours in **one day**.

Discussion

1 How long does it last?

a school lesson ______ minutes

b your sleep at night ______ hours

c recess time ______ minutes

d movie ______ hours

e golf tournament ______ days

f hockey game ______ minutes

2 What major festivals do you have in your city or your culture?
How long do they last?

____

____

Exercise 18

Complete:

a 1 day = ______ hours **b** 1 hour = ______ minutes

c In one day, the hour hand moves around the clock ______ times.

Exercise 19

There are 7 days in one **week**. Here are their names, written in order.

Practise your writing.

Sunday	Sunday	______
Monday	Monday	______
Tuesday	Tuesday	______
Wednesday	Wednesday	______
Thursday	Thursday	______
Friday	Friday	______
Saturday	Saturday	______

The days in the **weekend** are ______ and ______ .

Activity

Write down something you often do on:

a Sunday

____

____

b Monday

____

____

c Tuesday

____

____

d Wednesday

____

____

e Thursday

____

____

f Friday

____

____

g Saturday

____

____

Activity

Do you have a favourite day of the week?

______________________ is my favourite day because ______________________

____

Exercise 20

Write down the day *after*:

a Monday ______________________

b Thursday ______________________

c Saturday ______________________

Exercise 21

Write down the day *before*:

a Sunday ______________________

b Wednesday ______________________

c Monday ______________________

Exercise 22

There are 12 **months** in one **year**. Here are their names, written in order.

Practise your writing.

January	January	______
February	February	______
March	March	______
April	April	______
May	May	______
June	June	______
July	July	______
August	August	______
September	September	______
October	October	______
November	November	______
December	December	______

Exercise 23

Write down the month *after*:

a May ______

b July ______

c November ______

Exercise 24

Write down the month *before*:

a February ______

b April ______

c October ______

Activity

I was born in ______________________ .

My teacher was born in ______________________ .

Discussion

What **seasons** do you experience?

☐ I do not experience seasons. The weather is the same all year.

☐ I experience a wet season and a dry season.

The wet season is usually from

______________________ until ______________________ .

The dry season is usually from

______________________ until ______________________ .

☐ I experience 4 seasons each year. The months in each season are:

Summer	**Autumn or Fall**
____________________ ____________________ ____________________	____________________ ____________________ ____________________
Winter	**Spring**
____________________ ____________________ ____________________	____________________ ____________________ ____________________

CHAPTER 11: MONEY

Exercise 1

To make sure you are ready to work with money, practise your counting.

a Add 1 each time:

1 → 2 → ☐ → ☐ → 5 → ☐ → ☐ → 8

b Add 10 each time:

10 → 20 → ☐ → 40 → ☐ → 60 → ☐ → ☐

c Add 2 each time:

2 → 4 → ☐ → 8 → ☐ → ☐ → 14 → ☐

d Add 20 each time:

20 → ☐ → 60 → ☐ → ☐

e Add 5 each time:

5 → 10 → ☐ → ☐ → 25 → ☐ → ☐ → ☐

We use **money** to buy:

- food from a supermarket
- clothes from a department store
- tools from a hardware store.

Discussion

Where would you go to buy:

- a tennis ball ______________________
- a shirt ______________________
- bread ______________________
- a hammer? ______________________

Activity

Describe a recent visit to a shop. What type of shop was it? What did you buy when you were there?

....

....

....

....

In most countries we can use **coins** and **notes** to buy items.

coins

notes

Exercise 2

Practise your writing.

a money

money

....

....

b coins

coins

....

....

c notes

notes

....

....

Activity

a What are the rabbits doing in this picture? Can you explain why?

....

....

....

b Colour in the rest of the picture.

Different types of money are used in different countries.

Activity

You will need: one of each coin used in your country

What to do:

Carefully place a coin *under* the page.

Use a pencil to "rub" over the coin.

Rub *both sides* of each coin.

Discussion

What *colours* are the coins used in your country?

Click on the icon to print some Exercises for the notes and coins used in your country.

CHAPTER 12: CHANCE

Activity

a Describe something that **will happen** to you today.

Draw a picture of it.

b Describe something that **might happen** to you today.

Draw a picture of it.

c Describe something that **will not happen** to you today.

Draw a picture of it.

Activity

What things will happen to you next week? Use ✓ to choose.

	Will happen	*Might happen*	*Will not happen*
You will drink something.			
You will visit the doctor.			
You will visit the moon.			
You will trip over.			
You will be eaten by a dinosaur.			
You will go for a walk.			
You will swim in your pyjamas.			
You will sleep.			
You will kick a ball.			
You will ride a cow.			

Discuss your answers as a class.

Did everyone give the same answers?

Discuss why some answers may be different.

Exercise 1

Look at the cartoon.

Circle the correct answer.

- Billy *will* play tennis tomorrow.
- Billy *might* play tennis tomorrow.
- Billy *will not* play tennis tomorrow.

If something will *definitely* happen, we say it is **certain**.

If something will *definitely not* happen, we say it is **impossible**.

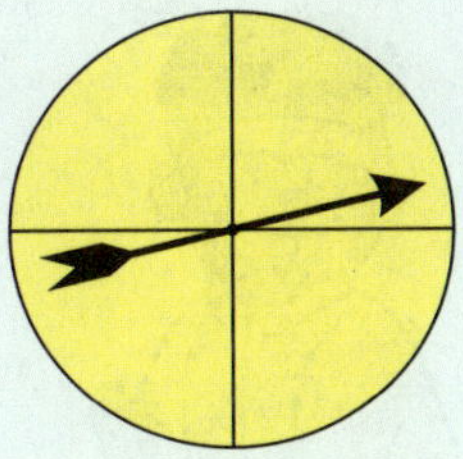

Exercise 2

Practise your writing.

a certain **b** maybe **c** impossible

certain maybe impossible

______ ______ ______

______ ______ ______

Exercise 3

Nima must select one of these three cards.

Complete each sentence using *impossible*, *possible*, or *certain*:

a It is ______ that Nima will select a car.

b It is ______ for Nima to select a bicycle.

c It is ______ that Nima will select the green car.

Exercise 4

Each child must select *one* ball from their bowl.

Complete each sentence using the correct name:

a It is certain that ____________ will select a red ball.

b It is impossible for ____________ to select a red ball.

c ____________ might select a red ball.

d ____________ might select a green ball.

e It is impossible for Amy and ____________ to select balls of the same colour.

Discussion

Anne spins this spinner once.

Which two colours *might* Anne spin?

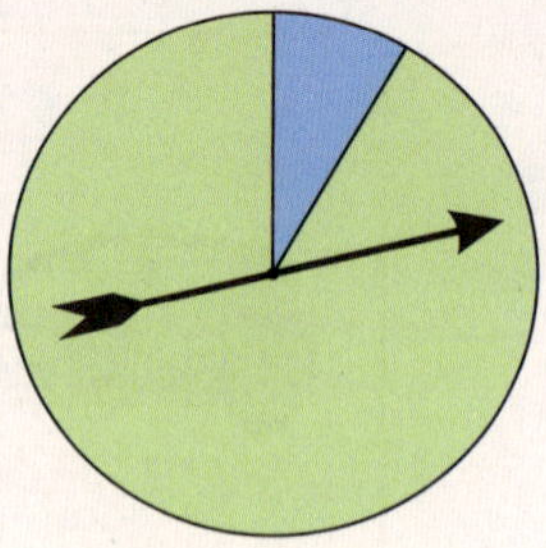

____________ ____________

Which colour do you *think* Anne will spin?

____

Explain your answer.

____

____

Revision — Number

1 Practise your writing.

1	one	11	______	30	______
2	______	12	______	40	______
3	______	13	______	50	______
4	______	14	______	60	______
5	______	15	______	70	______
6	______	16	______	80	______
7	______	17	______	90	______
8	______	18	______	100	______ ______
9	______	19	______		
10	______	20	______		

2 What number is shown?

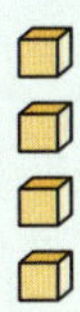

4
four

a 

______

______

b 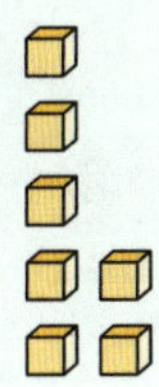

______

______

3 Count:

a 

______ frogs

b 

______ birds

4 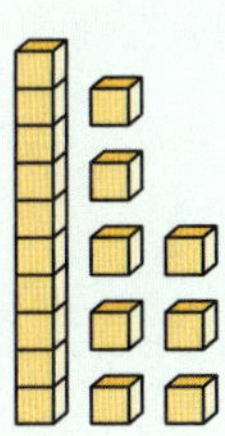

18 has ______ ten block and ______ unit blocks.

5 Draw eight circles.

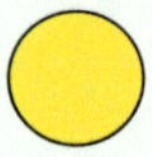

6 Use $=$ or $\neq$.

a 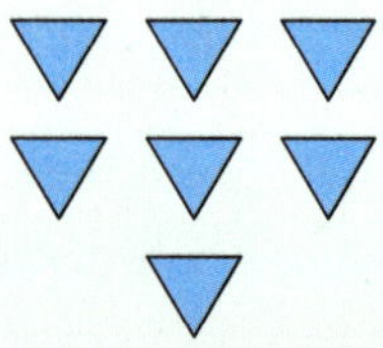 ______ 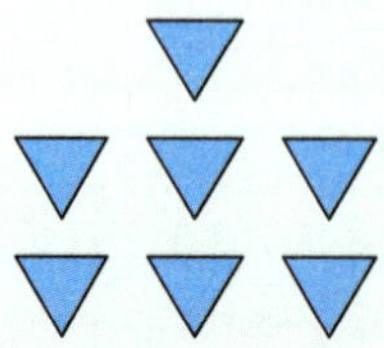

b 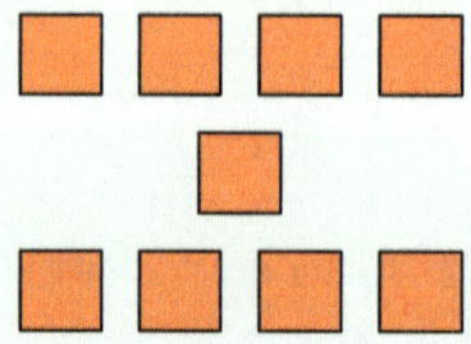 ______ 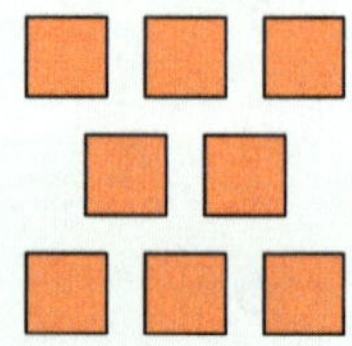

7 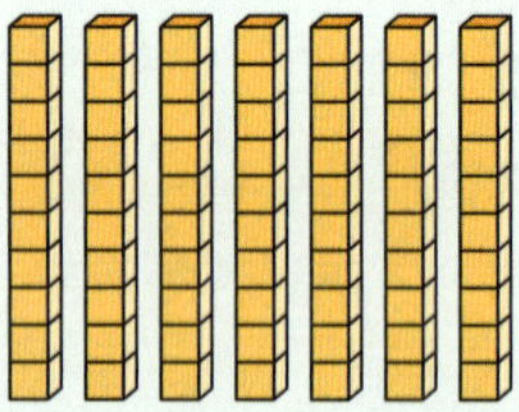

☐ tens + ☐ units = ☐

The number is ______________ .

There are ______________ units.

8 What numbers are shown?

a 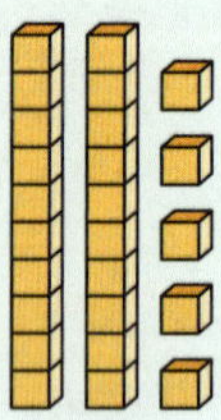

☐ tens and ☐ units

☐ + ☐ = ☐

____

b 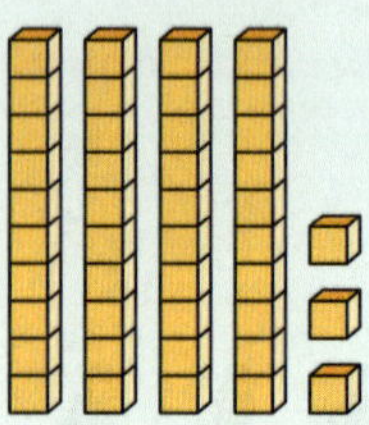

☐ tens and ☐ units

☐ + ☐ = ☐

____

9 Partition each number using place values:

a 3 9

☐ tens ☐ units

☐ + ☐

____

b 6 7

☐ tens ☐ units

☐ + ☐

____

10 Write each number in words:

a 42 ______________________

b 54 ______________________

c 71 ______________________

d 87 ______________________

e 96 ______________________

11 Use *more* or *fewer*.

a

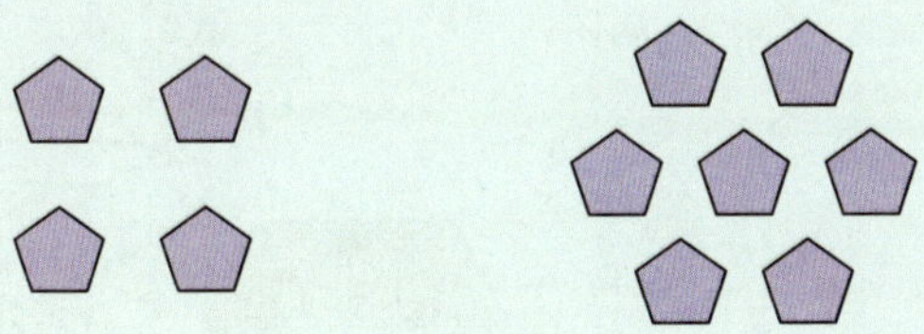

Four is ______________ than seven.

b

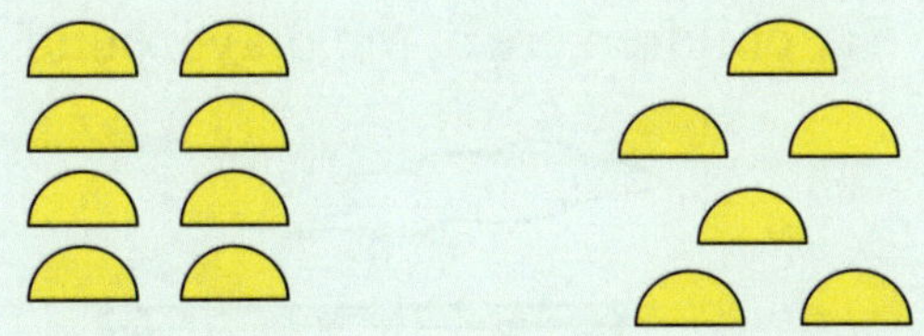

Eight is ______________ than six.

12 Use $>$ or $<$.

a 2 ____ 4 **b** 5 ____ 1 **c** 3 ____ 9

13 Complete each number line:

a

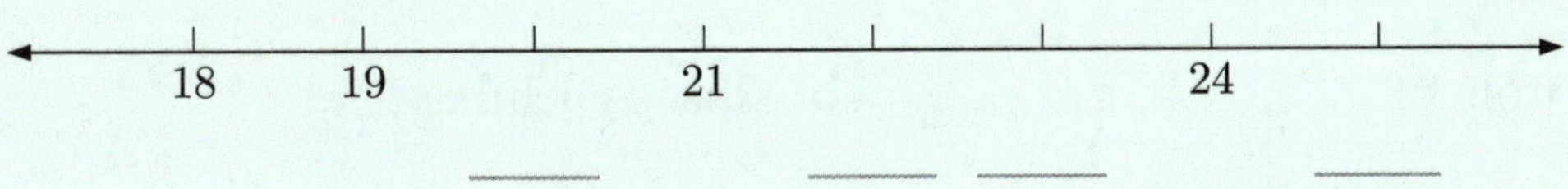

b

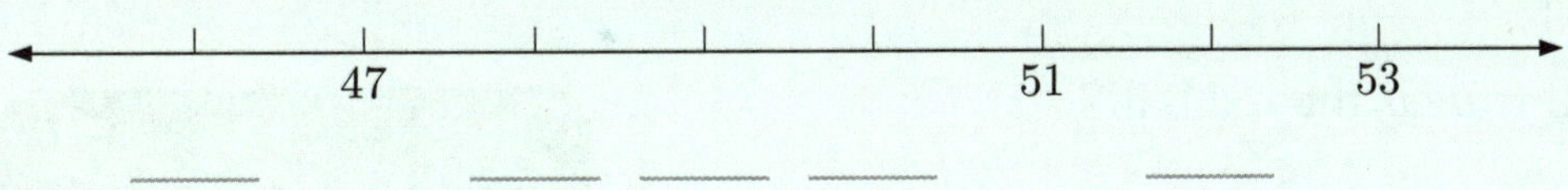

14

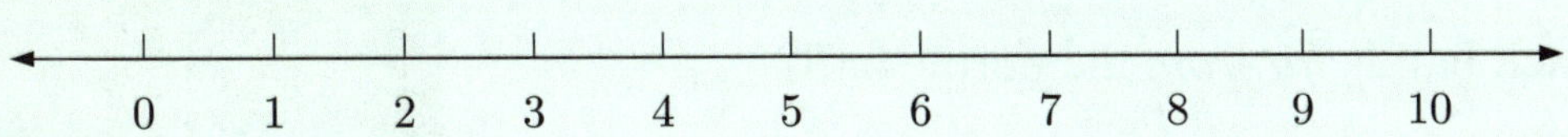

Circle the numbers *greater than* 4 in *red*.

Circle the numbers *less than* 4 in *blue*.

Revision — Position and direction

1 Complete each sentence using *left*, *right*, *above*, or *below*:

a The sea horse is to the .... of the dolphin.

b The helicopter is .... the turtle.

c The turtle is to the .... of the boat.

d The sea horse is .... the boat.

2 Six children are at the playground.

a Mila is in front of ....

b Elias is behind ....

c Ivy is alongside ....

3 **a** Which ball is *near* the red ball?

....

b Which ball is *far from* the purple ball?

....

c Draw a black ball *alongside* the white ball.

Revision — Addition

1 Use the number line to help you add:

a $3 + 2 =$ ______

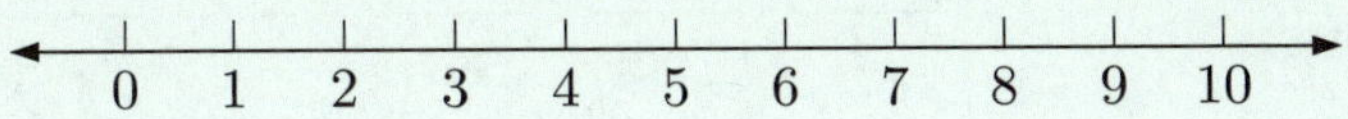

b $4 + 5 =$ ______

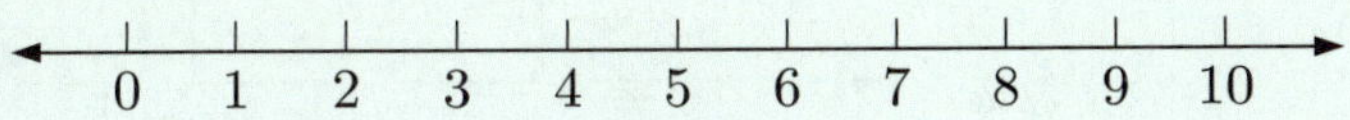

2 Colour objects to match the sum.

a $6 + 2 = 8$

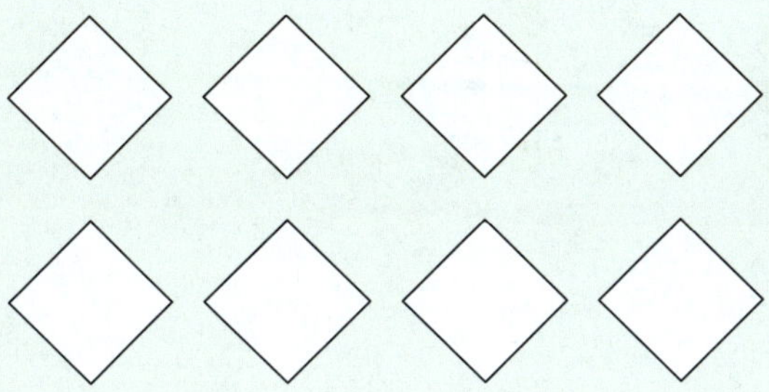

b $3 + 7 = 10$

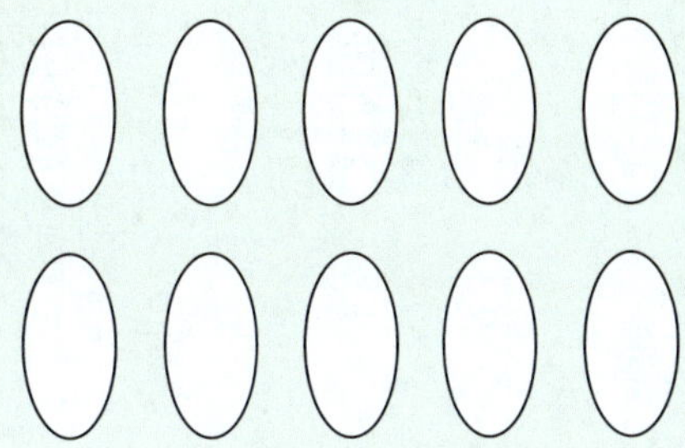

3 Find the sum:

a $2 + 7 =$ ______ **b** 1 add 6 = ______ **c** 5 plus 3 = ______

d 4 add 3 = ______ **e** $3 + 5 =$ ______ **f** 6 add 2 = ______

4 Write pairs of numbers that total 6.

$6 = 6 + 0$ $6 = \square + \square$ $6 = \square + \square$

$6 = \square + \square$ $6 = \square + \square$ $6 = \square + \square$

$6 = 0 + \square$

5 Use the number line to help you add:

a $3 + 9 =$ ______

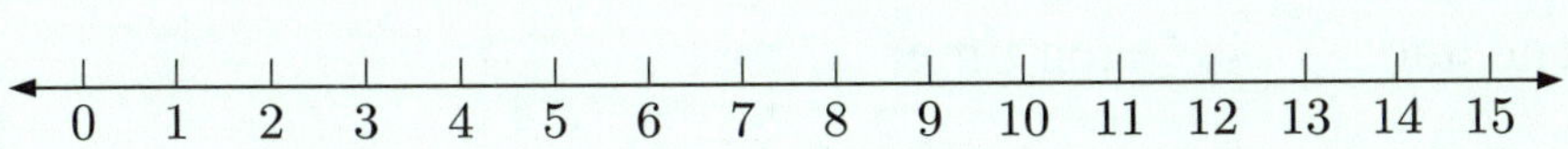

b $6 + 8 =$ ______

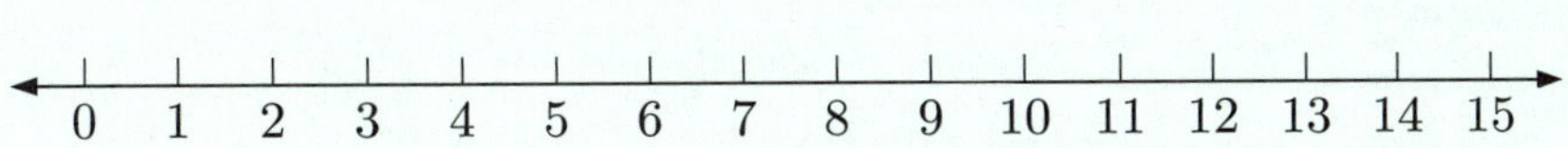

6 Colour objects to match the sum.

a $8 + 5 = 13$

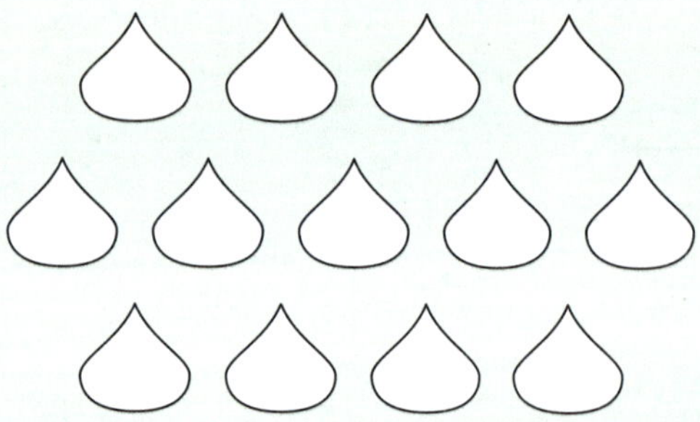

b $6 + 9 = 15$

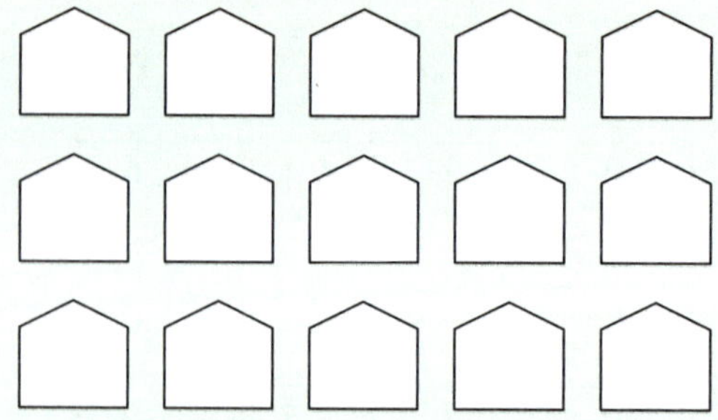

7 Complete:

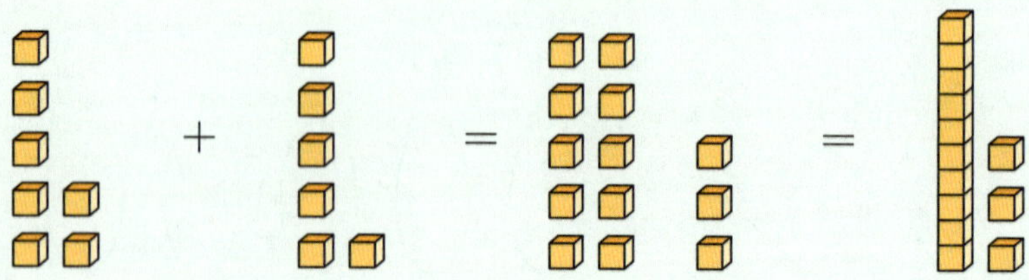

____ + ____ = 10 + ____ = ____

8 Use blocks to help you add:

a $4 + 7 =$ ____ **b** $6 + 6 =$ ____ **c** $9 + 5 =$ ____

9

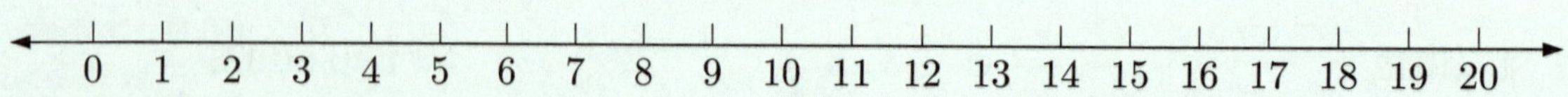

a Add 3 each time:

5 → ☐ → ☐ → ☐ → ☐

b Add 4 each time:

1 → ☐ → ☐ → ☐ → ☐

10 Describe the sequence:

2 → 4 → 6 → 8 → 10

I start with ____, and add ____ each time.

11 Complete these doubles:

a "Double 7" is $7 + 7 =$ ____.

b "Double 8" is ____ + ____ = ____.

12 Complete these pairs of additions.

a $7+0=$ ______ **b** $6+3=$ ______ **c** $8+4=$ ______

$0+7=$ ______ $3+6=$ ______ $4+8=$ ______

13 Complete these additions.

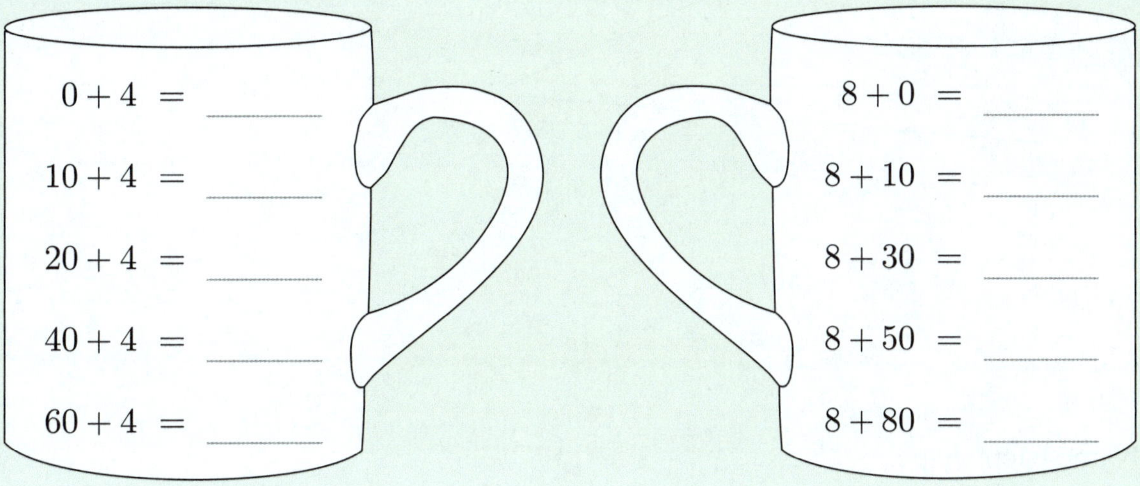

14 Complete these additions.

a $2+$ ______ $=7$ **b** $8+$ ______ $=11$ **c** ______ $+5=12$

15

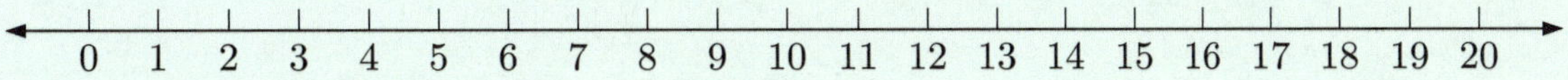

Complete these additions.

a $1+4+3=$ ______ **b** $5+2+6=$ ______ **c** $3+4+8=$ ______

16 Write an addition to solve each problem.

a Sam has 8 books and 5 magazines.

Altogether, Sam has ______________ = ______ items.

b Emma has 3 red dresses, 5 white dresses, and 2 green dresses.

In total, Emma has ______________ = ______ dresses.

17 Complete these additions.

a $3+4=$ ______ **b** $8+2=$ ______

$30+40=$ ______ $80+20=$ ______

Revision Shape

1 Fill in the labels using *edge*, *curve*, *straight line*, *triangle*, and *corner*.

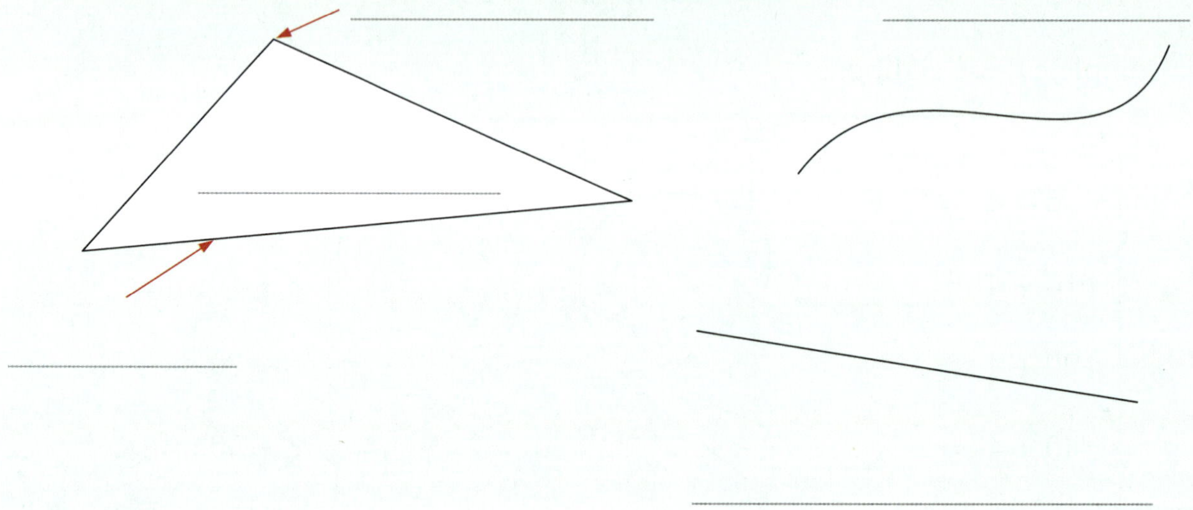

2 Name each shape:

a

b

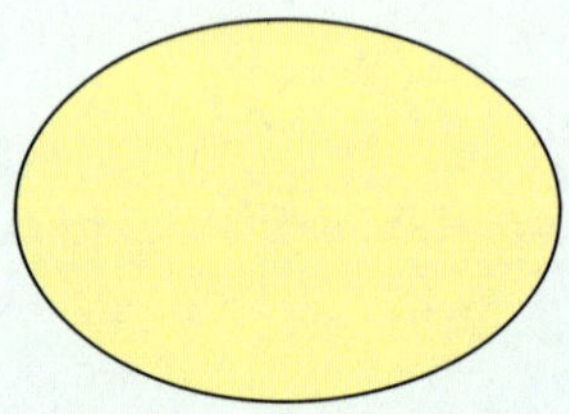

c

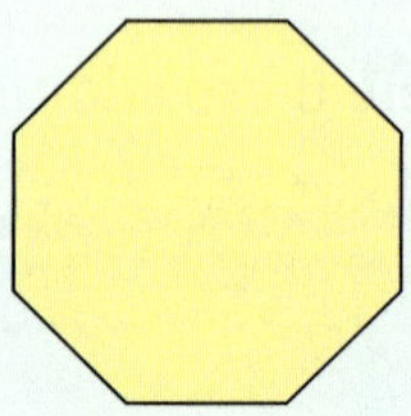

d

e

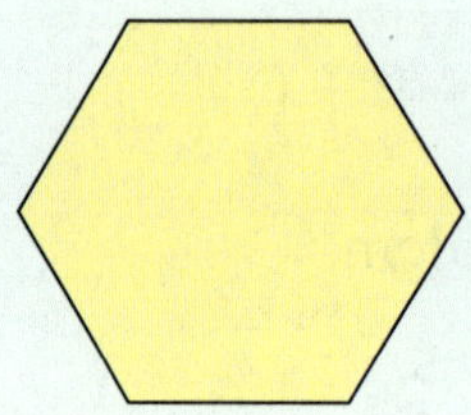

f

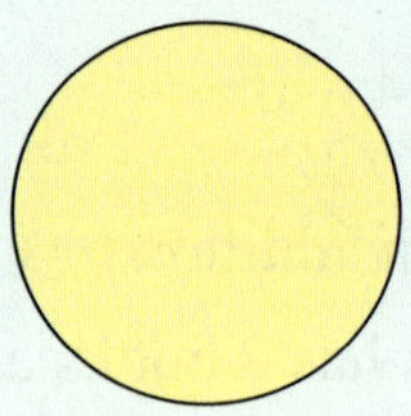

3 Colour the circles in purple, the squares in red, and the hexagons in yellow.

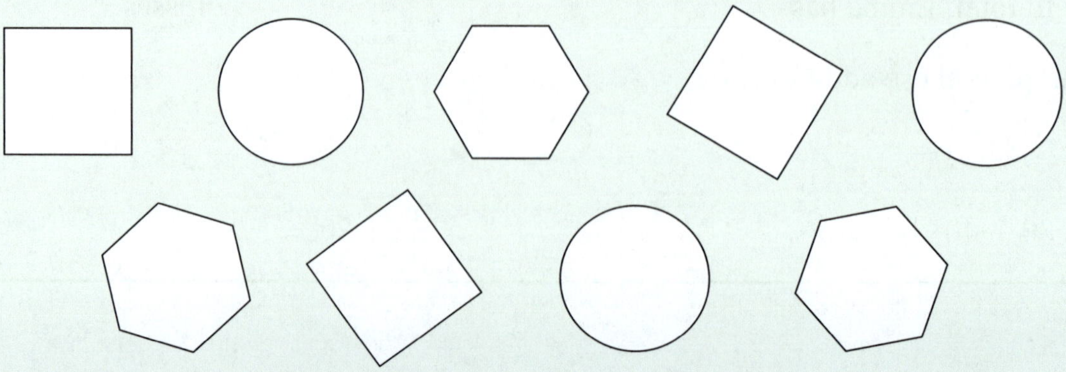

4 Complete:

a A triangle has ______ sides and ______ corners.

The sides are all ________________.

b A rectangle has ______ sides and ______ corners.

c An __________________ has 8 sides and ______ corners.

d An oval is drawn with a ________________.

5 Name each solid:

a

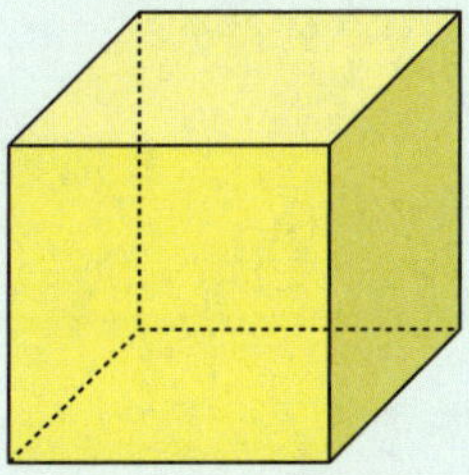

____

b

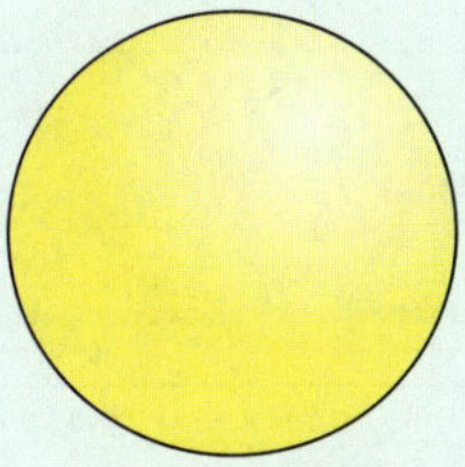

____

c

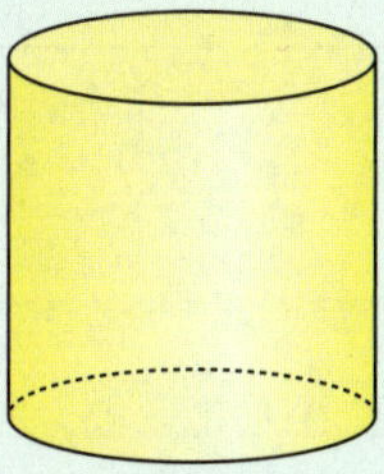

____

6 Complete:

a This is a ____________________ __________.

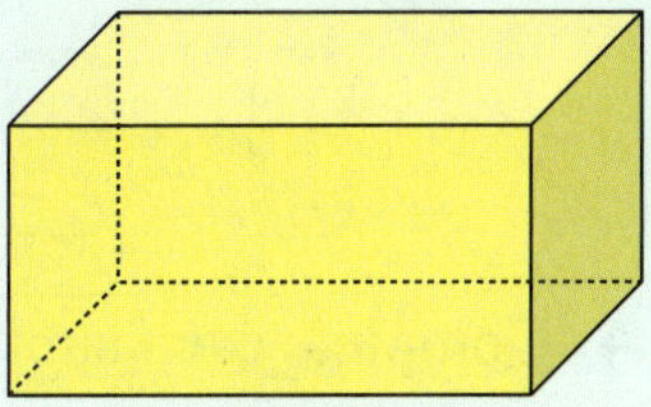

It has ______ faces, ______ edges,

and ______ corners.

b This is a __________________________.

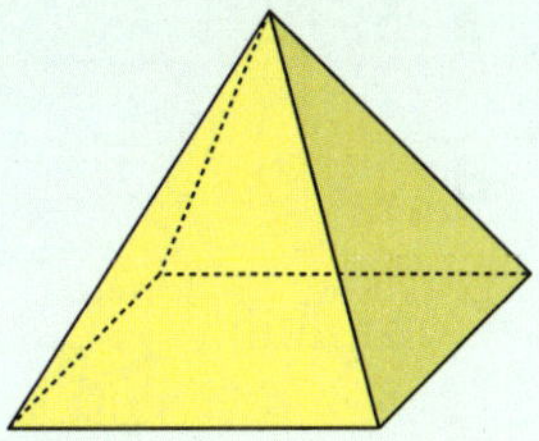

It has ______ faces, ______ edges,

and ______ corners.

One of its faces is a ________________.

Four of its faces are ________________.

7 **a** I am a solid with 6 faces. All of my faces are squares.

I am a ____________________.

b I am a solid. I have two flat surfaces and one curved surface.

I am a ____________________.

Revision — Subtraction

1 Use the pictures to write a subtraction:

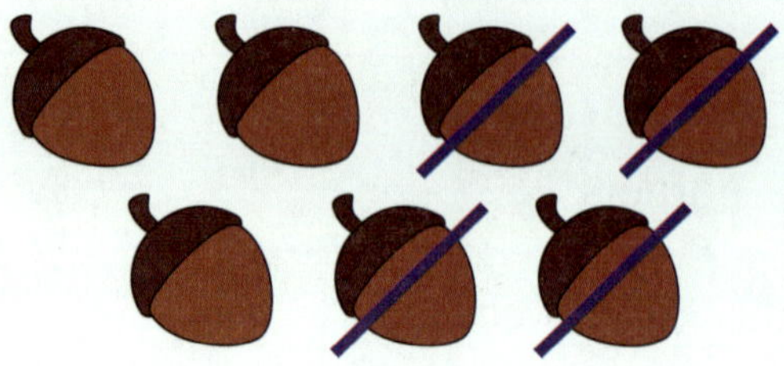

☐ – ☐ = ☐

2 Count backwards on the number line to help you subtract:

a $6 - 2 =$ ______

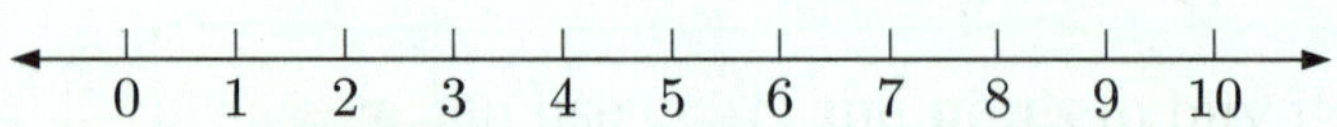

b $9 - 7 =$ ______

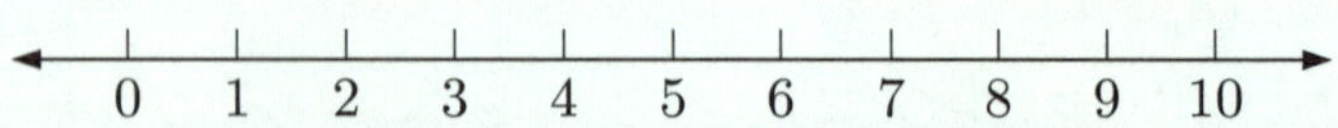

3 Cross out pictures to help you subtract:

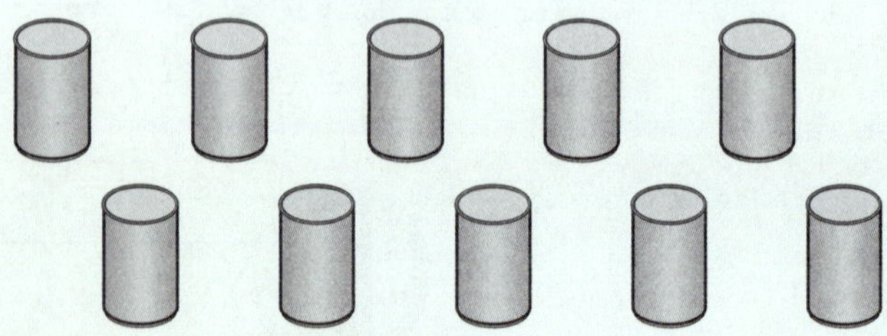

$10 - 8 =$ ______

4 Complete each subtraction:

a $5 - 1 =$ ______ **b** $7 - 3 =$ ______ **c** $9 - 4 =$ ______

d $4 - 2 =$ ______ **e** $8 - 7 =$ ______ **f** $6 - 4 =$ ______

5 Complete:

a 7 take away 5 = ______ **b** 8 minus 4 = ______

c 10 subtract 2 = ______

6 **a** Subtract 5 from 9. ______ – ______ = ______

b Take 4 away from 10. ______ – ______ = ______

7 Eric has 8 strawberries. He gives 6 to his mother.

______ – ______ = ______

Eric has ______ strawberries left.

8 Complete these subtractions.

a $2 - 0 =$ ______ **b** $8 - 8 =$ ______ **c** $6 - 0 =$ ______

9 Complete:

a	**b**
$4 - 4 =$ ______	$7 - 0 =$ ______
$5 - 4 =$ ______	$7 - 1 =$ ______
$6 - 4 =$ ______	$7 - 2 =$ ______
$7 - 4 =$ ______	$7 - 3 =$ ______
$8 - 4 =$ ______	$7 - 4 =$ ______
$9 - 4 =$ ______	$7 - 5 =$ ______
$10 - 4 =$ ______	$7 - 6 =$ ______
$11 - 4 =$ ______	$7 - 7 =$ ______

10

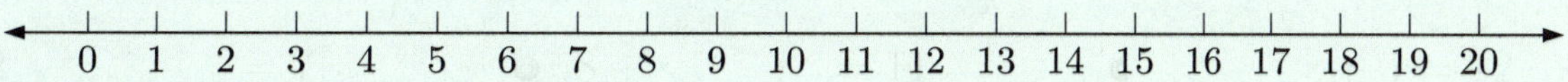

Use the number line to help you subtract:

a $12 - 6 =$ ______ **b** $15 - 5 =$ ______ **c** $18 - 3 =$ ______

d $16 - 7 =$ ______ **e** $14 - 8 =$ ______ **f** $13 - 4 =$ ______

11 Use the addition to write *two* subtractions.

a $4 + 8 = 12$

so $12 -$ ______ $=$ ______

and $12 -$ ______ $=$ ______

b $7 + 6 = 13$

so $13 -$ ______ $=$ ______

and $13 -$ ______ $=$ ______

12 Complete:

$11 - 10 =$ ______	$61 - 10 =$ ______
$21 - 10 =$ ______	$71 - 10 =$ ______
$31 - 10 =$ ______	$81 - 10 =$ ______
$41 - 10 =$ ______	$91 - 10 =$ ______
$51 - 10 =$ ______	

13 Complete:

a $25 - 10 =$ ______ **b** $37 - 10 =$ ______ **c** $64 - 10 =$ ______

14 **a** Take away 2 each time:

17 → ☐ → ☐ → ☐ → ☐ → ☐

b Take away 3 each time:

15 → ☐ → ☐ → ☐ → ☐ → ☐

15 Describe each sequence:

a 19 → 16 → 13 → 10 → 7

I start with ______, and ______________ ______ each time.

b 20 → 16 → 12 → 8 → 4

I start with ______, and ______________ ______ each time.

16 Complete:

a $7 - \square = 2$ **b** $12 - \square = 6$ **c** $14 - \square = 5$

d $\square - 4 = 5$ **e** $\square - 6 = 4$ **f** $\square - 5 = 8$

17 Complete these subtractions.

a $50 - 10 =$ ______ **b** $80 - 10 =$ ______

$50 - 20 =$ ______ $80 - 20 =$ ______

$50 - 30 =$ ______ $80 - 30 =$ ______

$50 - 40 =$ ______ $80 - 40 =$ ______

$50 - 50 =$ ______ $80 - 50 =$ ______

Revision — Measurement

1

shark  whale 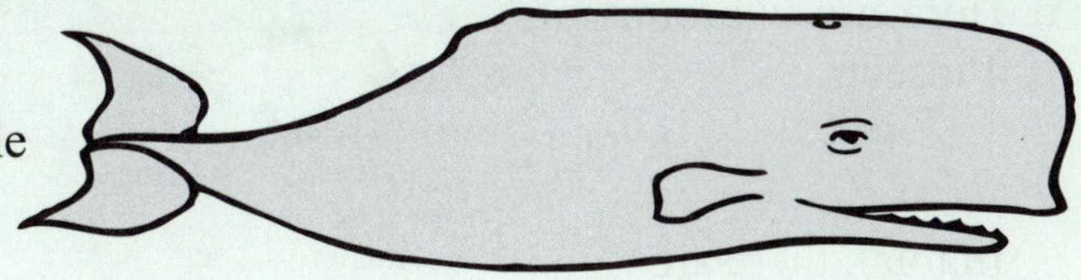

tuna  salmon 

Complete using *longer*, *shorter*, *longest*, or *shortest*.

a The salmon is ____________________.

b The whale is ____________________.

c The shark is ____________________ than the salmon.

d The tuna is ____________________ than the shark.

2 **a** The smallest square is ______.

b The largest square is ______.

c Square **A** is larger than square ______.

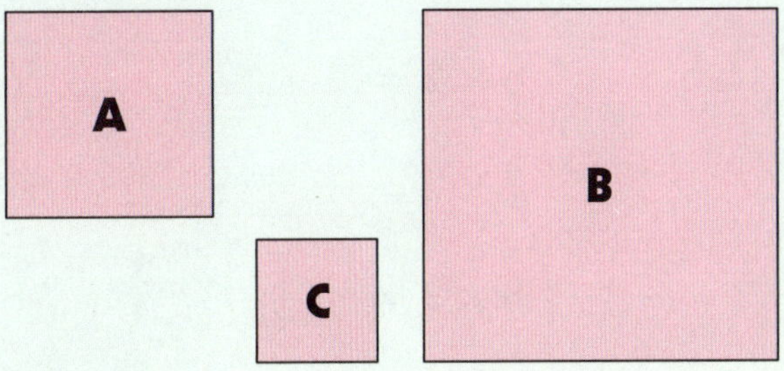

3 Look at these containers.

The ____________________ has the smallest capacity.

The ____________________ has the largest capacity.

vase   water bottle   milk carton

4 Complete using *less than*, *more than*, or *the same as*.

a The shell weighs

____________________ 2 blocks.

b The rock weighs

____________________ 5 blocks.

Revision — Multiplication

1 Here are 3 groups of 4 insects.

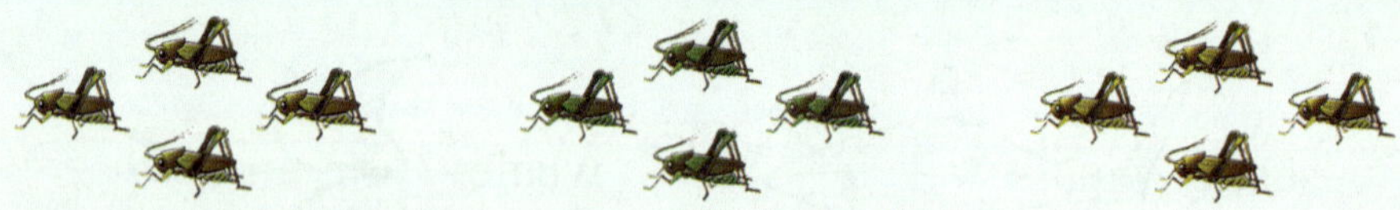

In total, there are ______ + ______ + ______ = ______ insects.

As a multiplication, we write ______ $\times 4 =$ ______

2 Complete these times tables.

$1 \times 2 =$ ______	$1 \times 3 =$ ______	$1 \times 5 =$ ______
$2 \times 2 =$ ______	$2 \times 3 =$ ______	$2 \times 5 =$ ______
$3 \times 2 =$ ______	$3 \times 3 =$ ______	$3 \times 5 =$ ______
$4 \times 2 =$ ______	$4 \times 3 =$ ______	$4 \times 5 =$ ______
$5 \times 2 =$ ______	$5 \times 3 =$ ______	$5 \times 5 =$ ______
$6 \times 2 =$ ______	$6 \times 3 =$ ______	$6 \times 5 =$ ______
$7 \times 2 =$ ______	$7 \times 3 =$ ______	$7 \times 5 =$ ______
$8 \times 2 =$ ______	$8 \times 3 =$ ______	$8 \times 5 =$ ______
$9 \times 2 =$ ______	$9 \times 3 =$ ______	$9 \times 5 =$ ______
$10 \times 2 =$ ______	$10 \times 3 =$ ______	$10 \times 5 =$ ______

3 Complete this table:

Picture	*Addition*	*Multiplication*	*Answer*
	$2+2+2$		
		5×3	

4 Complete these times tables.

$1 \times 1 =$ ______

$2 \times 1 =$ ______

$3 \times 1 =$ ______

$4 \times 1 =$ ______

$5 \times 1 =$ ______

$6 \times 1 =$ ______

$7 \times 1 =$ ______

$8 \times 1 =$ ______

$9 \times 1 =$ ______

$10 \times 1 =$ ______

$1 \times 10 =$ ______

$2 \times 10 =$ ______

$3 \times 10 =$ ______

$4 \times 10 =$ ______

$5 \times 10 =$ ______

$6 \times 10 =$ ______

$7 \times 10 =$ ______

$8 \times 10 =$ ______

$9 \times 10 =$ ______

$10 \times 10 =$ ______

5 **a** There are 6 rows with 2 dots in each row.

Complete:

$2 + 2 + 2 + 2 + 2 + 2 =$ ______

$6 \times 2 =$ ______

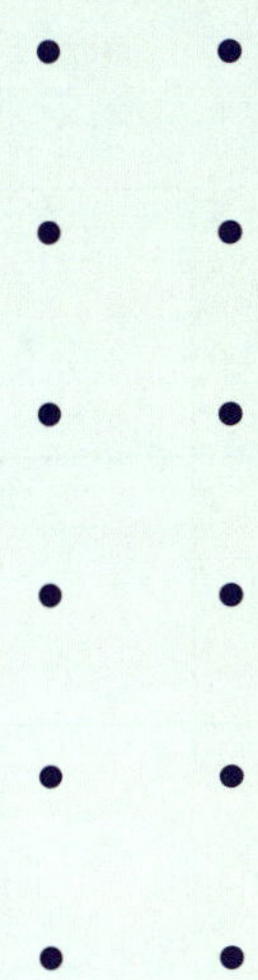

b There are 2 columns with 6 dots in each column.

Complete:

$6 + 6 =$ ______

$2 \times 6 =$ ______

Revision — Data handling

1 Count the fruit using a tally.

Fruit	*Tally*
peach	
pear	
plum	

There are ☐ peaches, ☐ pears, and ☐ plums.

There are ☐ fruit in total.

2 Henry counted the fish in an aquarium.

Draw a pictograph to show this information.

Use 🐟 = 1 fish

Fish

Goldfish	3
Clownfish	4
Minnow	7
Guppy	5

Goldfish	
Clownfish	
Minnow	
Guppy	

Revision — Division

1 Describe the division:

a

______ has been divided into

______ equal groups of ______ .

b

______ has been divided into

______ equal groups of ______ .

2 Shade one half of each shape:

a

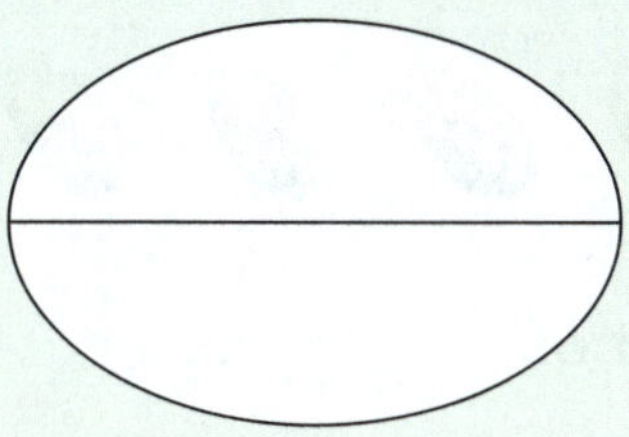

b

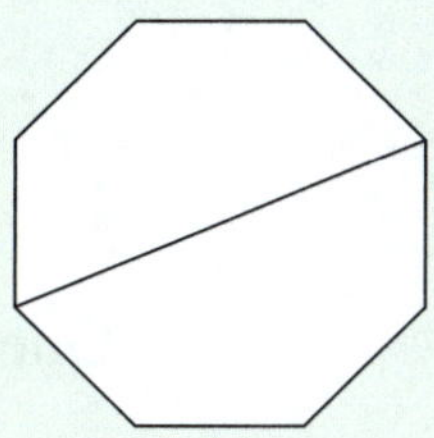

c

3 Circle one half of the objects.

a

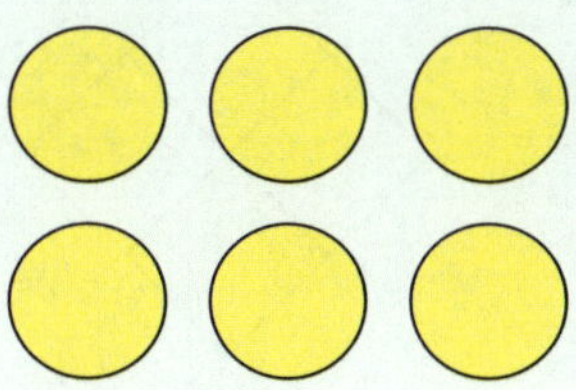

One half of ______ is ______ .

b

One half of ______ is ______ .

4 Shade one quarter of each shape:

a

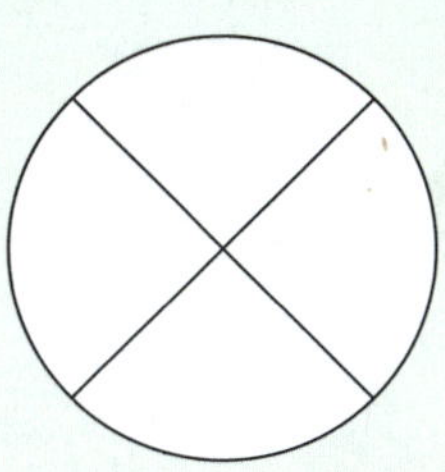

b

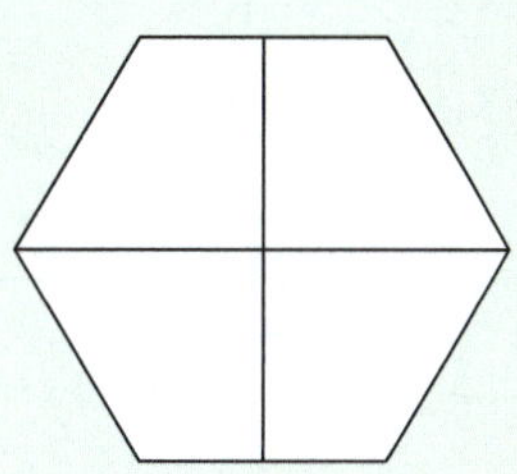

c

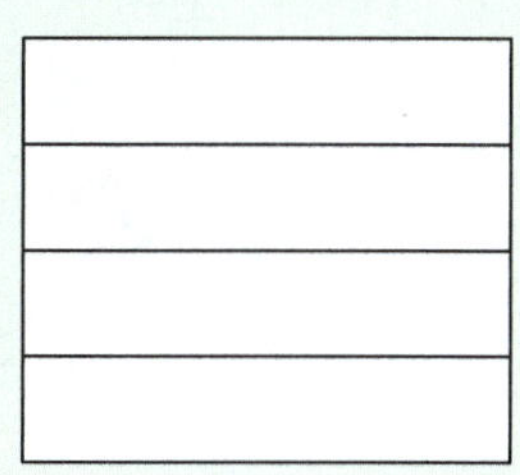

5 Shade the correct amount.

a one quarter

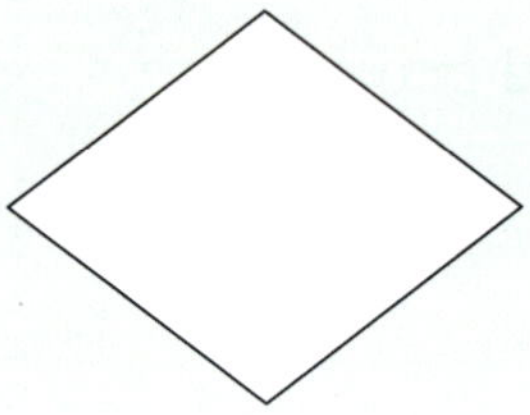

b two quarters

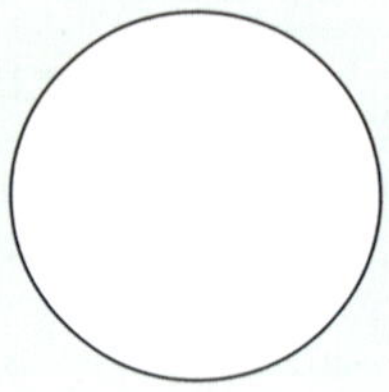

c three quarters

6 Circle one quarter of the objects.

a

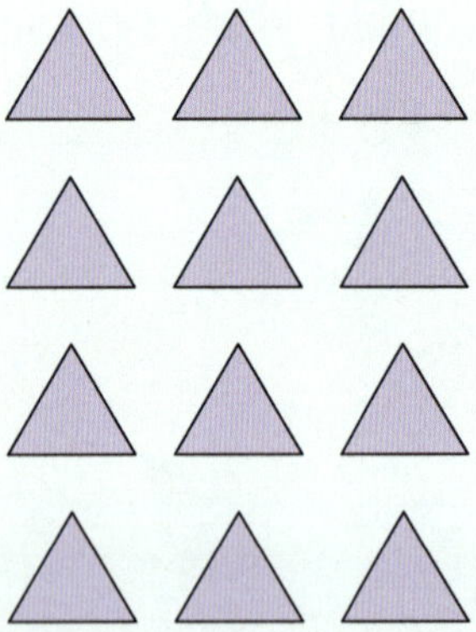

One quarter of ______ is ______ .

b

One quarter of ______ is ______ .

7 How much is shaded?

a

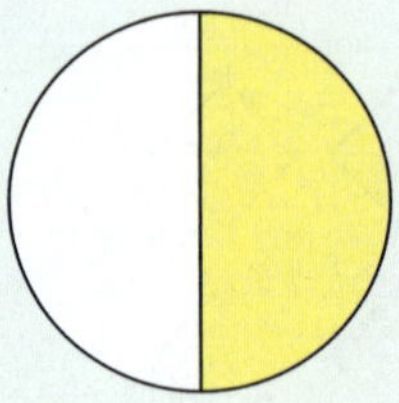

____

b

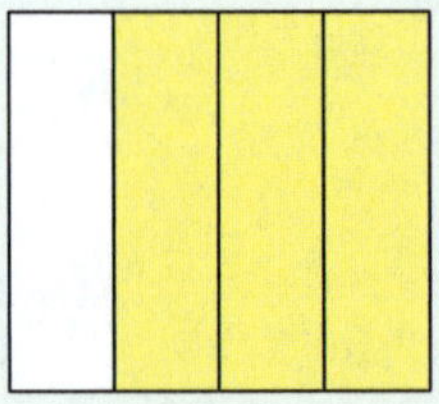

____

c

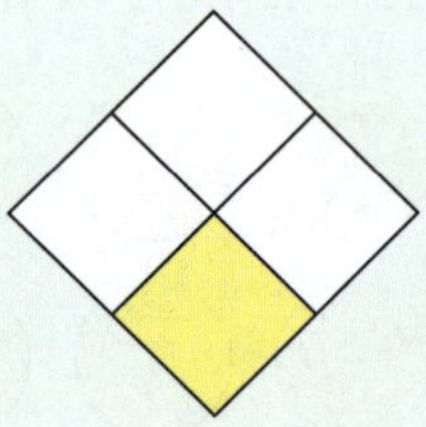

____

Revision — Time

1 Circle what you do *last* each day.

go to school   go to bed   eat lunch

2 What time is shown on the clock?

a

b
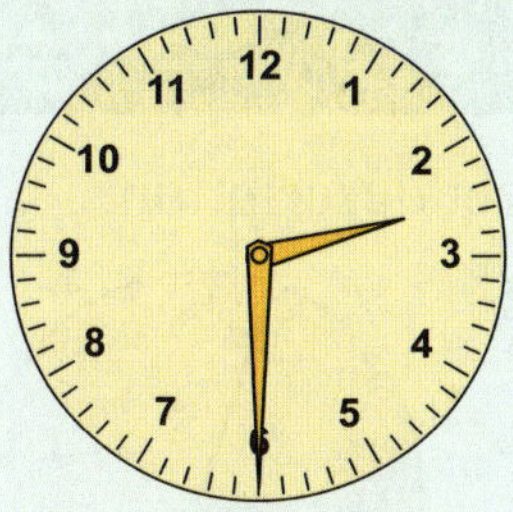

c

d
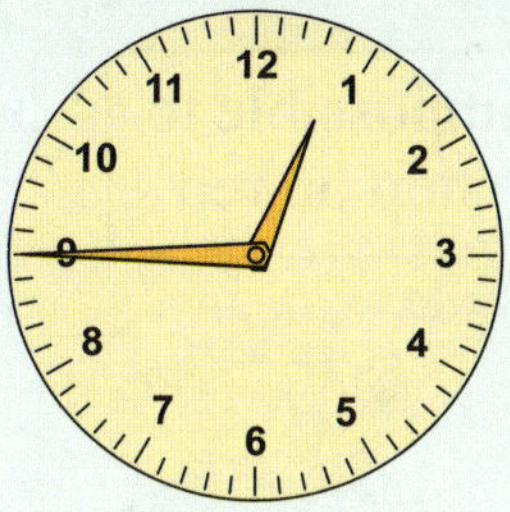

e
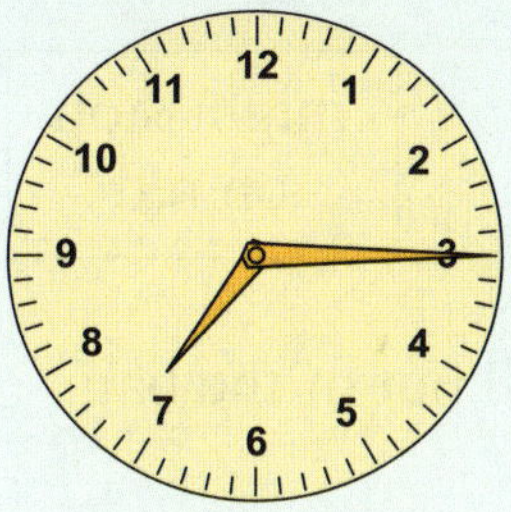

f
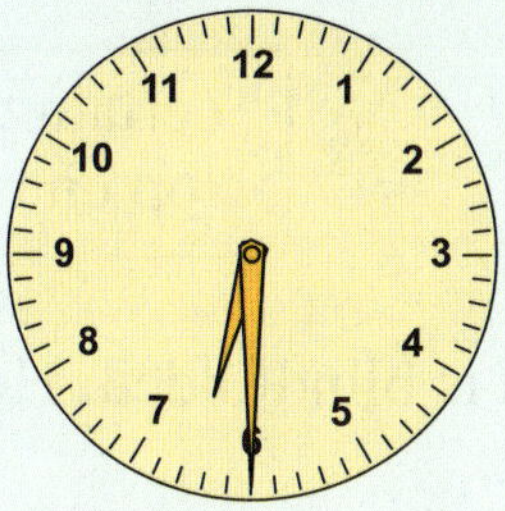

3 Draw the hands on each clock face to show the time:

a half past 4
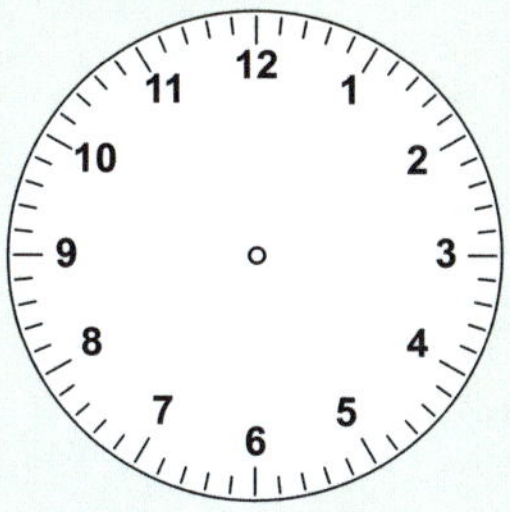

b 8 o'clock
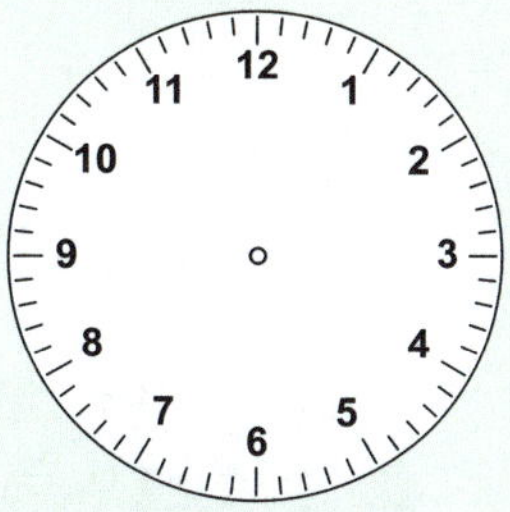

c half past 9
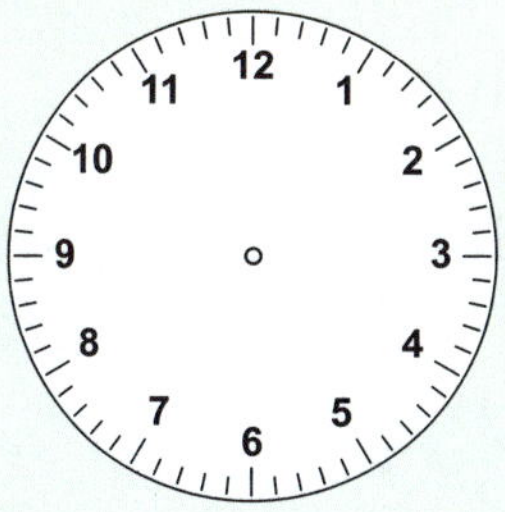

4 There are:

a ________ minutes in 1 hour

b ________ hours in 1 day

c ________ days in 1 week

d ________ months in 1 year.

5 **a** The day after Wednesday is ____________________.

b The day before Tuesday is ____________________.

6 a The month after August is ______________________ .

b The month before June is ______________________ .

Revision — Chance

1 Match each spinner with the correct description.

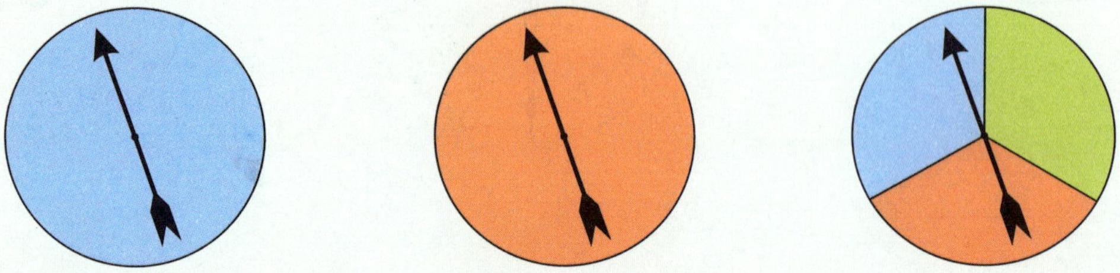

certain to stop on red	might stop on red	impossible to stop on red

2 Ted is blindfolded. He picks a ball out of the bag.

Complete each sentence using *impossible*, *possible*, or *certain*:

a It is ______________________ that Ted will pick a blue ball.

b It is ______________________ that Ted will pick the number 3.

c It is ______________________ for Ted to pick a green ball.